BIORTHOGONALITY
AND ITS APPLICATIONS
TO NUMERICAL ANALYSIS

MONOGRAPHS AND TEXTBOOKS IN
PURE AND APPLIED MATHEMATICS

82. *T. Petrie and J. D. Randall,* Transformation Groups on Manifolds (1984)
83. *K. Goebel and S. Reich,* Uniform Convexity, Hyperbolic Geometry, and Non-expansive Mappings (1984)
84. *T. Albu and C. Năstăsescu,* Relative Finiteness in Module Theory (1984)
85. *K. Hrbacek and T. Jech,* Introduction to Set Theory: Second Edition, Revised and Expanded (1984)
86. *F. Van Oystaeyen and A. Verschoren,* Relative Invariants of Rings: The Non-commutative Theory (1984)
87. *B. R. McDonald,* Linear Algebra Over Commutative Rings (1984)
88. *M. Namba,* Geometry of Projective Algebraic Curves (1984)
89. *G. F. Webb,* Theory of Nonlinear Age-Dependent Population Dynamics (1985)
90. *M. R. Bremner, R. V. Moody, and J. Patera,* Tables of Dominant Weight Multiplicities for Representations of Simple Lie Algebras (1985)
91. *A. E. Fekete,* Real Linear Algebra (1985)
92. *S. B. Chae,* Holomorphy and Calculus in Normed Spaces (1985)
93. *A. J. Jerri,* Introduction to Integral Equations with Applications (1985)
94. *G. Karpilovsky,* Projective Representations of Finite Groups (1985)
95. *L. Narici and E. Beckenstein,* Topological Vector Spaces (1985)
96. *J. Weeks,* The Shape of Space: How to Visualize Surfaces and Three-Dimensional Manifolds (1985)
97. *P. R. Gribik and K. O. Kortanek,* Extremal Methods of Operations Research (1985)
98. *J.-A. Chao and W. A. Woyczynski, eds.,* Probability Theory and Harmonic Analysis (1986)
99. *G. D. Crown, M. H. Fenrick, and R. J. Valenza,* Abstract Algebra (1986)
100. *J. H. Carruth, J. A. Hildebrant, and R. J. Koch,* The Theory of Topological Semigroups, Volume 2 (1986)
101. *R. S. Doran and V. A. Belfi,* Characterizations of C*-Algebras: The Gelfand-Naimark Theorems (1986)
102. *M. W. Jeter,* Mathematical Programming: An Introduction to Optimization (1986)
103. *M. Altman,* A Unified Theory of Nonlinear Operator and Evolution Equations with Applications: A New Approach to Nonlinear Partial Differential Equations (1986)
104. *A. Verschoren,* Relative Invariants of Sheaves (1987)
105. *R. A. Usmani,* Applied Linear Algebra (1987)
106. *P. Blass and J. Lang,* Zariski Surfaces and Differential Equations in Characteristic p > 0 (1987)
107. *J. A. Reneke, R. E. Fennell, and R. B. Minton,* Structured Hereditary Systems (1987)
108. *H. Busemann and B. B. Phadke,* Spaces with Distinguished Geodesics (1987)
109. *R. Harte,* Invertibility and Singularity for Bounded Linear Operators (1988)
110. *G. S. Ladde, V. Lakshmikantham, and B. G. Zhang,* Oscillation Theory of Differential Equations with Deviating Arguments (1987)
111. *L. Dudkin, I. Rabinovich, and I. Vakhutinsky,* Iterative Aggregation Theory: Mathematical Methods of Coordinating Detailed and Aggregate Problems in Large Control Systems (1987)
112. *T. Okubo, Differential Geometry* (1987)
113. *D. L. Stancl and M. L. Stancl,* Real Analysis with Point-Set Topology (1987)
114. *T. C. Gard,* Introduction to Stochastic Differential Equations (1988)
115. *S. S. Abhyankar,* Enumerative Combinatorics of Young Tableaux (1988)
116. *H. Strade and R. Farnsteiner,* Modular Lie Algebras and Their Representations (1988)
117. *J. A. Huckaba,* Commutative Rings with Zero Divisors (1988)
118. *W. D. Wallis,* Combinatorial Designs (1988)
119. *W. Więsław,* Topological Fields (1988)

120. *G. Karpilovsky,* Field Theory: Classical Foundations and Multiplicative Groups (1988)
121. *S. Caenepeel and F. Van Oystaeyen,* Brauer Groups and the Cohomology of Graded Rings (1989)
122. *W. Kozlowski,* Modular Function Spaces (1988)
123. *E. Lowen-Colebunders,* Function Classes of Cauchy Continuous Maps (1989)
124. *M. Pavel,* Fundamentals of Pattern Recognition (1989)
125. *V. Lakshmikantham, S. Leela, and A. A. Martynyuk,* Stability Analysis of Nonlinear Systems (1989)
126. *R. Sivaramakrishnan,* The Classical Theory of Arithmetic Functions (1989)
127. *N. A. Watson,* Parabolic Equations on an Infinite Strip (1989)
128. *K. J. Hastings,* Introduction to the Mathematics of Operations Research (1989)
129. *B. Fine,* Algebraic Theory of the Bianchi Groups (1989)
130. *D. N. Dikranjan, I. R. Prodanov, and L. N. Stoyanov,* Topological Groups: Characters, Dualities, and Minimal Group Topologies (1989)
131. *J. C. Morgan II,* Point Set Theory (1990)
132. *P. Biler and A. Witkowski,* Problems in Mathematical Analysis (1990)
133. *H. J. Sussmann,* Nonlinear Controllability and Optimal Control (1990)
134. *J.-P. Florens, M. Mouchart, and J. M. Rolin,* Elements of Bayesian Statistics (1990)
135. *N. Shell,* Topological Fields and Near Valuations (1990)
136. *B. F. Doolin and C. F. Martin,* Introduction to Differential Geometry for Engineers (1990)
137. *S. S. Holland, Jr.,* Applied Analysis by the Hilbert Space Method (1990)
138. *J. Okniński,* Semigroup Algebras (1990)
139. *K. Zhu,* Operator Theory in Function Spaces (1990)
140. *G. B. Price,* An Introduction to Multicomplex Spaces and Functions (1991)
141. *R. B. Darst,* Introduction to Linear Programming: Applications and Extensions (1991)
142. *P. L. Sachdev,* Nonlinear Ordinary Differential Equations and Their Applications (1991)
143. *T. Husain,* Orthogonal Schauder Bases (1991)
144. *J. Foran,* Fundamentals of Real Analysis (1991)
145. *W. C. Brown,* Matrices and Vector Spaces (1991)
146. *M. M. Rao and Z. D. Ren,* Theory of Orlicz Spaces (1991)
147. *J. S. Golan and T. Head,* Modules and the Structures of Rings: A Primer (1991)
148. *C. Small,* Arithmetic of Finite Fields (1991)
149. *K. Yang,* Complex Algebraic Geometry: An Introduction to Curves and Surfaces (1991)
150. *D. G. Hoffman, D. A. Leonard, C. C. Lindner, K. T. Phelps, C. A. Rodger, and J. R. Wall,* Coding Theory: The Essentials (1991)
151. *M. O. González,* Classical Complex Analysis (1992)
152. *M. O. González,* Complex Analysis: Selected Topics (1992)
153. *L. W. Baggett,* Functional Analysis: A Primer (1992)
154. *M. Sniedovich,* Dynamic Programming (1992)
155. *R. P. Agarwal,* Difference Equations and Inequalities (1992)
155. *C. Brezinski,* Biorthogonality and Its Applications to Numerical Analysis (1992)

Additional Volumes in Preparation

BIORTHOGONALITY AND ITS APPLICATIONS TO NUMERICAL ANALYSIS

Claude Brezinski

Université des Sciences et Techniques de Lille Flandres–Artois
Lille Flandres–Artois, France

Marcel Dekker, Inc.   **New York • Basel • Hong Kong**

ISBN 0-8247-8616-5

This book is printed on acid-free paper.

MARCEL DEKKER, INC.
270 Madison Avenue, New York, New York 10016

Current printing (last digit):
10 9 8 7 6 5 4 3 2 1

PRINTED IN THE UNITED STATES OF AMERICA

Preface

The solution of the general interpolation problem has very many applications in numerical analysis and applied mathematics. However, some time ago, I realized that its possibilities have not been fully exploited and have even been underestimated, and I began to work on the subject. The concept underlying the problem is that of biorthogonality which gave its title to this book. It has many unusual connections and applications to Fourier expansion, projections, divided differences, extrapolation processes, numerical methods for integrating differential equations or for solving integral equations, rational approximations to formal power series and series of functions, least squares, statistics, and biorthogonal polynomials, to name some.

Most of the results given in this book are new and have not even been published in the form of journal articles. They appear here for the first time. This is the case in particular for the various recurrence relations given and for the generalizations of the method of moments, the method of Lanczos, and the biconjugate gradient method. New approximations of Padé-type for series are also described.

The possibilities opened by the concept of biorthogonality have still to be discovered and new applications as well. Thus, this book will be of interest to researchers in numerical analysis and approximation theory. However, this does not mean that the material given here is difficult. Almost no prerequisite are needed and the book can also be used as a text for students.

I hope that this monograph will be useful to many applied mathematicians and will serve as a basis for new developments and applications.

I would like to thank Professor Zuhair Nashed and Professor Earl Taft, who accepted the book in their series. I express my gratitude to Professor Jet Wimp for his encouragement during the preparation of the manuscript. My thanks are also due to Mrs. Françoise Tailly who carefully typed the manuscript, and Ms. Maria Allegra of Marcel Dekker, Inc., for their assistance in the production of the book.

Claude Brezinski

Contents

BIORTHOGONALITY
AND ITS APPLICATIONS
TO NUMERICAL ANALYSIS

1 - INTRODUCTION

Numerical analysis is concerned with the operator equation

$$Af = b$$

where $f \in E$, $b \in F$, E and F are vector spaces and $A : E \rightarrow F$. Three different problems can be treated. They are, by increasing order of difficulty [123] :

- the direct problem : given A and f, compute b (for example : the computation of definite integrals)
- the inverse problem : given A and b, compute f (for example : the resolution of systems of equations)
- the identification problem : given f and b, compute A (for example : the approximation of functions).

When E and F are infinite dimensional spaces, the solution of the preceding problems is, in general, impossible. Even when E and/or F are finite dimensional spaces, their solutions can pose serious difficulties. In these cases, the initial problem is replaced by an approximate one in finite dimensional spaces (or in spaces with fewer dimensions)

$$A_n f_n = b_n$$

where $f_n \in E_n$, $b_n \in F_n$, E_n and F_n are vector spaces of finite dimensions and $A_n : E_n \rightarrow F_n$.

This approximate problem is called the discretization of the original problem and the main question is to measure the distance between the exact solution of the original problem and the exact solution of the discretized problem (the discretization error) and to study the convergence when the dimensions of E_n and F_n tend to infinity.

Sometimes, even the discretized problem cannot be solved exactly as is the case for systems of nonlinear equations. The method used cannot lead to its exact solution and we have a method's error which must be also studied.

Finally when using a computer, we are faced to rounding errors due to the computer's arithmetic.

Of course, the study of these errors needs that the vector spaces are normed and the study of convergence requires that they are normed and complete that is that they are Banach spaces. Thus the tools and the methods of functional analysis play a central rôle in numerical analysis

1

which was first emphasized by L.V. Kantorovich in 1948. These questions were discussed by many authors, for example [50, 123].

However, the first step is to replace the original problem by the discretized one. It turns out that many numerical methods used for this purpose can be reformulated in the framework of projection methods. Such methods form a very broad class of methods, as stated by Cryer [50], especially if they are looked as generalized collocation methods as in [154].

Our main interest here will be on algorithms for constructing discretized problems by generalized collocation methods. We shall make use of the old concept of biorthogonality which can be traced back to the book of Banach [6] or even before since the special case of biorthogonal systems of functions can be found, for example, in the treatise of analysis of Goursat [82] but goes back to the works of Hilbert and others on Fredholm's integral equations between 1904 and 1910. It seems that this concept has not yet been fully developed and exploited although a renewal of interest in biorthogonal polynomials has been recently observed (see [110] and the references quoted therein). Since the problem we shall be treating is an algebraic one, we shall not make use of topology and thus the spaces we shall be dealing with will be vector spaces unless specified.

2 - PRELIMINARIES

Let E be a vector space on K (R or $\mathbb{C}$) and E^* its dual (the vector space of linear functionals on E).

Let us first recall some classical results whose proofs can be found in [57].

Theorem 1. *Let* E_n *be a subspace of dimension* $n+1$ *of* E. *If* $x_0, x_1, ..., x_n$ *are linearly independent in* E_n *and if* $L_0, L_1, ..., L_n$ *are independent in* E^* *then the determinant*

$$G_{n+1} = |L_i(x_j)|_{i,j=0}^{n} \neq 0.$$

Conversely, if either $x_0, ..., x_n$, *or* $L_0, ..., L_n$ *are independent and if* $G_{n+1} \neq 0$, *then the other set is also independent.*

Since the polynomial case will be important in the sequel, let us illustrate the preceding theorem and the following ones by such an example.

Let E be the space of functions defined on D $\subset \mathbb{C}$ and let $E_n = P_n$ the vector space of polynomials of degree at most n. $x_i = z^i$ are independent in P_n. The functionals L_i are defined by $\forall f \in E$, $L_i(f) = f(z_i)$ where $z_i \in D$. G_{n+1} is a Vandermonde determinant which is different from zero if and only if $\forall i \neq j$, $z_i \neq z_j$.

We shall now have a look at the interpolation problem and begin with an existence and uniqueness result.

Theorem 2 : *Let* E_n *be a subspace of dimension* n+1 *of* E, *let* $x_0,...,x_n$ *be independent in* E_n *and let* $L_0,...,L_n$ *belong to* E^*. *The general interpolation problem : find* $R_n \in E_n$ *such that* $L_i(R_n) = w_i$ *for* i = 0,...,n *has a unique solution for arbitrary values of* $w_0,...,w_n$ *not all zero, if and only if* $L_0,...,L_n$ *are independent in* E^*.

Coming back to the polynomial case this is the well known result stating the existence and uniqueness of the interpolation polynomial under the necessary and sufficient condition that all the interpolation points are distinct.

The solution of the general interpolation problem, as stated in theorem 2, can be expressed in a determinantal form

Theorem 3 : *Under the assumptions of theorem 2, the solution* R_n *of the general interpolation problem is given by*

$$R_n = - \begin{vmatrix} 0 & x_0 & ... & x_n \\ w_0 & L_0(x_0) & ... & L_0(x_n) \\ ... & ... & ... & ... \\ w_n & L_n(x_0) & ... & L_n(x_n) \end{vmatrix} \Big/ G_{n+1}$$

where G_{n+1} *is defined as in theorem 1 and where the determinant in the numerator of* R_n *denotes the linear combination of the elements in its first row obtained by the classical rule for expanding a determinant.*

In the polynomial case this is the well-known expression of the interpolation polynomial as a ratio ot two determinants. Such a representation is not suitable for practical computations since the computation of a determinant requires too many arithmetical operations (k.k! multiplications for a determinant of order k). A more convenient representation is given by the following theorem

Theorem 4 : *Under the assumptions of theorem 2, there are* $n+1$ *uniquely determined independent elements of* E_n, *denoted by* $x'_o,...,x'_n$, *such that*

$$L_i(x'_j) = \delta_{ij}.$$

$\forall f \in E_n$ *we have*

$$f = \sum_{i=o}^{n} L_i(f) \, x'_i.$$

For every choice of $w_o,...,w_n$, *the unique solution* R_n *of the general interpolation problem is given by*

$$R_n = \sum_{i=o}^{n} w_i \, x'_i.$$

In the polynomial case, when $L_i(f) = f(z_i)$ it can be proved that

$$x'_i = \prod_{\substack{j=o \\ j \neq i}}^{n} \frac{z - z_j}{z_i - z_j}$$

and the above formula is Lagrange's representation of the interpolation polynomial.

In the preceding formula $\forall i$, x'_i is a linear combination of $x_o,...,x_n$ thus leading to the main drawback of Lagrange's formula : if we want to increase n, we must determine an entirely new set of elements $y'_o,...,y'_{n+1}$ which are not simply related to the old ones $x'_o,...,x'_n$. In the polynomial case the remedy is classical : it is Newton's formula which consists in constructing simultaneously two new basis, $L_0^*,...,L_n^*$ and $x_0^*,...,x_n^*$, such that $L_i^*(x_j^*) = \delta_{ij}$ but with the difference that L_i^* and x_i^* are now linear combinations of only $L_0,...,L_i$ and $x_0,...,x_i$ respectively instead of the whole set. This remedy enables us to solve the interpolation problem recursively that is just by adding one new term when passing from n to $n+1$, a property known as the permanence property of Newton's representation (which is also characteristic of Fourier expansions).

The same trick can be used for the general interpolation problem via the concept of biorthogonal family which will be now studied.

3 - BIORTHOGONALITY AND APPLICATIONS

The notion of biorthogonality is obviously a generalization of the notion of orthogonality in an Hilbert space which itself comes from the notion of orthogonality for functions and polynomials. Chapter VII of Banach's book of 1932 is devoted to the general notion of biorthogonality. Although biorthogonality received some attention since that time, it was only quite recently that the study of biorthogonal polynomials in connection with some problems in rational approximation and numerical methods for ordinary differential equations, appeared on the scene and played a central rôle (see, for example, [110]). Orthogonality of dimension d for polynomials [103] and, equivalently, 1/d-orthogonality [131] were recently the subjects of investigations and applications. All these new notions of orthogonality for polynomials are particular cases of the general notion of biorthogonality which also provides, as we shall see below, a natural and general framework for the definition and the study of generalizations of many concepts and methods such as the methods of moments and that of Galerkin, Lanczos' bi-orthogonalization process, the bi-conjugate gradient method, projections, Padé approximants of various types, extrapolation methods for scalar and vector sequences, and so on.

Let us now give the general setting of biorthogonality as explained by Davis [57]

Theorem 5 : *Let* E *be an infinite dimensional vector space. Let* $x_0, x_1, \ldots$ *be a sequence of elements of* E *such that* $\forall n$, $x_0, \ldots, x_n$ *are linearly independent. Let* $L_0, L_1, \ldots$ *be a sequence of linear functionals in* E^* *such that* $\forall n$, $G_{n+1} \neq 0$.
Then there are uniquely determined constants a_{ij} *and* b_{ij}, *with* $a_{ii} \neq 0$ *such that*

$$L_0^* = a_{00}L_0 \qquad\qquad x_0^* = x_0$$
$$L_1^* = a_{10}L_0 + a_{11}L_1 \qquad\qquad x_1^* = b_{10}x_0 + x_1$$
$$L_2^* = a_{20}L_0 + a_{21}L_1 + a_{22}L_2 \qquad\qquad x_2^* = b_{20}x_0 + b_{21}x_1 + x_2$$

- - - - - - - - - - - - - - - - -   - - - - - - - - - - - - - - -

with

$$\overset{*}{L_i}(x_j) = \delta_{ij}.$$

We have

$$\overset{*}{x_i} = \begin{vmatrix} L_0(x_0) & ... & L_0(x_i) \\ ... & ... & ... \\ L_{i-1}(x_0) & ... & L_{i-1}(x_i) \\ x_0 & ... & x_i \end{vmatrix} / G_i$$

$$\overset{*}{L_i} = \begin{vmatrix} L_0(x_0) & ... & L_i(x_0) \\ ... & ... & ... \\ L_0(x_{i-1}) & ... & L_i(x_{i-1}) \\ L_0 & ... & L_i \end{vmatrix} / G_{i+1}.$$

Let $E_n = \text{Span}(x_0,...,x_n)$. *Then*, $\forall f \in E_n$, $f = \sum_{i=o}^{n} \overset{*}{L_i}(f) \, \overset{*}{x_i}$.

Let $\overset{*}{E_n} = \text{Span}(L_0,...,L_n)$. *Then*, $\forall L \in \overset{*}{E_n}$, $L = \sum_{i=o}^{n} L(\overset{*}{x_i}) \overset{*}{L_i}$.

$\{\overset{*}{L_i}, \overset{*}{x_j}\}$ is called a biorthogonal family.

From this result we see that the solution R_n of the general interpolation problem in E_n that is to find R_n such that

$$L_i(R_n) = L_i(f) \qquad\qquad \text{for} = i = 0,...,n$$

is given by the Newton's formula

$$R_n = \sum_{i=o}^{n} \overset{*}{L_i}(f) \overset{*}{x_i},$$

and we have

$$R_{n+1} = R_n + \overset{*}{L_{n+1}}(f) \overset{*}{x_{n+1}} \quad \text{with} \quad R_0 = \frac{L_0(f)}{L_0(x_0)} x_0.$$

We see that we also have

$$L_j^*(x_i) = L_i^*(x_j) = 0 \qquad\qquad \text{for } j = 0,\ldots,i\text{-}1$$

and that

$$L_i^*(x_i) = G_{i+1}/G_i$$
$$L_i^*(x_i) = 1.$$

It follows that

$$G_{n+1} = |L_i(x_j)|_{i,j=0}^{n} = |L_i^*(x_j)|_{i,j=0}^{n} = \prod_{i=0}^{n} L_i^*(x_i).$$

Of course, there is a strong connection between interpolation and biorthogonality. We have

$$x_i^* = \begin{vmatrix} x_i & x_0 & \cdots & x_{i\text{-}1} \\ L_0(x_i) & L_0(x_0) & \cdots & L_0(x_{i\text{-}1}) \\ \cdots & \cdots & \cdots & \cdots \\ L_{i\text{-}1}(x_i) & L_{i\text{-}1}(x_0) & \cdots & L_{i\text{-}1}(x_{i\text{-}1}) \end{vmatrix} / G_i$$

$$= \begin{vmatrix} 0 & x_0 & \cdots & x_{i\text{-}1} \\ L_0(x_i) & L_0(x_0) & \cdots & L_0(x_{i\text{-}1}) \\ \cdots & \cdots & \cdots & \cdots \\ L_{i\text{-}1}(x_i) & L_{i\text{-}1}(x_0) & \cdots & L_{i\text{-}1}(x_{i\text{-}1}) \end{vmatrix} / G_i + x_i.$$

Thus $R_{i\text{-}1} = x_i - x_i^*$ where $R_{i\text{-}1}$ satisfies the interpolation conditions

$$L_j(R_{i\text{-}1}) = L_j(x_i) \qquad\qquad \text{for } j = 0,\ldots,i\text{-}1$$

that is

$$L_j(x_i - x_i^*) = L_j(x_i) \qquad\qquad \text{for } j = 0,\ldots,i\text{-}1$$

or again

$$L_j(x_i^*) = 0 \qquad\qquad \text{for } j \leq i\text{-}1.$$

As is the case for polynomials we can also define quasi-biorthogonality : $\{\bar{L}_i, \bar{x}_j\}$ is said to be a quasi-biorthogonal family of order $(p,q) \in \mathbb{N}^2$ if and only if

$$\bar{L}_i, (\bar{x}_j) = 0 \qquad \text{for } p < i\text{-}j \text{ and } i\text{-}j < \text{-}q$$
$$\neq 0 \qquad \text{for } i - j = p \text{ and } i\text{-}j = \text{-}q.$$

Of course quasi-biorthogonality of order $(0,0)$ reduces to biorthogonality. Let us assume that

$$\bar{L}_i = a_{i0}\overset{*}{L}_0 + ... + a_{ii}\overset{*}{L}_i$$
$$\bar{x}_j = b_{j0}\overset{*}{x}_0 + ... + b_{jj}\overset{*}{x}_j.$$

We have

$$\overset{*}{L}_i(\bar{x}_j) = b_{j0}\overset{*}{L}_i(\overset{*}{x}_j) + ... + b_{jj}\overset{*}{L}_i(\overset{*}{x}_j).$$

The condition $\overset{*}{L}_i(\bar{x}_j) = 0$ for $i = 0,...,j\text{-}q\text{-}1$ implies $b_{j0} = ... = b_{j,j\text{-}q\text{-}1} = 0$. We also have

$$\bar{L}_i(\overset{*}{x}_j) = a_{i0}\overset{*}{L}_0(\overset{*}{x}_j) + ... + a_{ii}\overset{*}{L}_i(\overset{*}{x}_j)$$

and the condition $\bar{L}_i(\overset{*}{x}_j) = 0$ for $j = 0,....,i\text{-}p\text{-}1$ implies $a_{i0} = ... = a_{i,i\text{-}p\text{-}1} = 0$.

The conditions $\overset{*}{L}_i(\bar{x}_j) \neq 0$ for $i = j\text{-}q$ and $\bar{L}_i(\overset{*}{x}_j) \neq 0$ for $j = i\text{-}p$ also imply $b_{j,j\text{-}q} \neq 0$ and $a_{i,i\text{-}p} \neq 0$.
Thus we have

$$\bar{L}_i = a_{i,i\text{-}p}\overset{*}{L}_{i\text{-}p} + ... + a_{ii}\overset{*}{L}_i$$

$$\bar{x}_j = b_{j,j\text{-}q}\overset{*}{x}_{j\text{-}q} + ... + b_{jj}\overset{*}{x}_j$$

which shows that $\bar{L}_i$ and $\bar{x}_j$ depend respectively on $p+1$ and $q+1$ arbitrary coefficients.

We have

$$\bar{L_i}(\bar{x}_j) = a_{i,i-p}\overset{*}{L}_{i-p}(\bar{x}_j) + ... + a_{ii}\overset{*}{L_i}(\bar{x}_j)$$

$$= a_{i,i-p}b_{j,j-q} \overset{*}{L}_{i-p} (\overset{*}{x}_{j-q}) + ... + a_{i,i-p}b_{jj}\overset{*}{L}_{i-p}(\overset{*}{x}_j)$$

$$+ a_{i,i-p+1}b_{j,j-q} \overset{*}{L}_{i-p+1} (\overset{*}{x}_{j-q}) + ... + a_{i,i-p+1}b_{jj}\overset{*}{L}_{i-p+1}(\overset{*}{x}_j)$$

$$+ a_{ii}b_{j,j-q} \overset{*}{L_i}(\overset{*}{x}_{j-q}) + ... + a_{i,i}b_{jj}\overset{*}{L_i}(\overset{*}{x}_j).$$

Thus $\bar{L_i}(\bar{x}_j) = 0$ for $p < i-j$ and $i-j < -q$.

Moreover when $i-j = p$ we have

$$\bar{L_i}(\bar{x}_j) = a_{i,i-p}b_{jj}$$

which implies $b_{jj} \neq 0$. We also have when $i-j = -q$

$$\bar{L_i}(\bar{x}_j) = a_{ii}b_{j,j-q}$$

which means that $a_{ii} \neq 0$.

3.1 - Orthogonality for polynomials.

If $E = P$, the vector space of polynomials, and if the functionals L_i are defined by, $\forall p \in P$

$$L_i(p) = \int_a^b p(x)\, \omega(x, \mu_i)\, d\alpha(x)$$

then the $\overset{*}{x_i}$'s of theorem 5 are the so-called bi-orthogonal polynomials introduced by Iserles and Nørsett in [107]. If the functionals L_i are not necessarily defined by an integral but are only known by their moments $L_i(x^j) = c_{ij}$ for $j = 0,1,...$ we obtain the (formal) bi-orthogonal polynomials of [26] which generalize those of Iserles and Nørsett. In general these orthogonal polynomials do not satisfy a recurrence relationship.

Now if we assume that

$$L_{d+i}(x^j) = L_i (x^{j+1}) \qquad\qquad \text{for } i = 0,1,...$$

then

$$L_{md+i}(x^j) = L_i(x^{j+m}) \qquad\qquad \text{for } i = 0,...,d-1$$

and the bi-orthogonal polynomials obtained are the so-called orthogonal polynomials of dimension d defined and studied by van Iseghem in [103]. Such polynomials satisfy an order d+1 recurrence relationship that is a relation between d+2 consecutive polynomials. These polynomials are equivalent to the 1/d-orthogonal polynomials introduced by Maroni [131] in a different context.

Orthogonal polynomials of dimension d were introduced in the study of vector Padé approximants [101] which are Padé approximants approximating simultaneously d power series by rational functions with a common denominator. Another kind of simultaneous Padé approximants was introduced by de Bruin [41]. As pointed out in [112] their denominators are also related to bi-orthogonal polynomials.

When d = 1, the classical formal orthogonal polynomials satisfying the usual three terms recurrence relationship are recovered [17]. Such polynomials can be generalized to the case where E is a commutative algebra. Let c be a linear functional on E. The functionals L_i are defined by, $\forall f \in E$

$$L_i(f) = c(x_if).$$

Then

$$L_i(x_j) = c(x_ix_j) = c(x_jx_i) = L_j(x_i)$$

and we have

$$a_{ij} = b_{ij}G_i/G_{i+1}.$$

Thus

$$c(x_i^*x_j^*) = c(x_i(b_{j0}x_0 + \dots + b_{jj}^*x_j))$$

$$= b_{j0}\, c(x_ix_0) + \dots + b_{jj}\, c(x_i^*\ x_j)$$

$$= b_{j0}\, L_0(x_i^*) + \dots + b_{jj}\, L_j(x_i^*)$$

$$= \frac{G_{i+1}}{G_j}\, (a_{j0}L_0 + \dots + a_{jj}L_j)(x_i^*)$$

$$= \frac{G_{j+1}}{G_j} L_j^*(x_i^*) = \frac{G_{j+1}}{G_j} \delta_{ij}.$$

If the determinants G_j are all positive, let us set

$$\bar{x}_i = (G_i/G_{i+1})^{1/2} x_i^*.$$

Then

$$c(\bar{x}_i \ \bar{x}_j) = \delta_{ij}.$$

Orthogonal polynomials are also known to satisfy the Christoffel-Darboux identity which is proved from the three-terms recurrence relationship. An interesting open question was to know whether or not bi-orthogonal polynomials or vector orthogonal polynomials or those of de Bruin could satisfy the Christoffel-Darboux identity without satisfying the usual three-terms recurrence relationship. Thus the first step was to find a direct proof of the Christoffel-Darboux identity for the usual orthogonal polynomials not making use of the recurrence relationship. Such a proof is given in appendix 1. Then the second step was to try to extend this proof to more general orthogonal polynomials. This attempt failed since, as shown in appendix 1, if the Christoffel-Darboux identity holds for a family of polynomials then this family satisfies a three-terms recurrence relationship whose coefficients can be deduced from that of the Christoffel-Darboux identity and thus it is a usual orthogonal family. This result proves the equivalence between orthogonal polynomials, a three-terms recurrence relationship (by an extension to the formal case of Favard's theorem, [17, p. 155]) and the Christoffel-Darboux identity. However some kind of generalization will be given in section 3.3. Orthogonality on a curve can also be treated within this framework. We shall come back to orthogonal polynomials in section 5.4.

3.2 - Interpolation and projection.

Let $f \in E$. We already saw that the solution R_n of the general interpolation problem in E_n, that is to find $R_n \in E_n = \text{Span} (x_0,...,x_n)$ such that

$$L_i(R_n) = L_i(f) = w_i \qquad \qquad \text{for } i = 0,...,n$$

is given by Newton's formula

$$R_n = \sum_{i=o}^{n} L_i^*(f)\, x_i^*$$

thus leading to the recursive scheme

$$R_0 = \frac{L_0(f)}{L_0(x_0)}\, x_0$$

$$R_{n+1} = R_n + L_{n+1}^*(f)\, x_{n+1}^*.$$

Thus R_n has the form

$$R_n = \sum_{i=o}^{n} a_i x_i^*.$$

with $a_i = L_i^*(f)$.

Often, in practice, the L_i^*'s are much more difficult to obtain than the x_i^*'s. This is the reason why we shall now give another expression for the a_i's.

We have

$$R_n = a_0 x_0^* + ... + a_n x_n^*$$

and the interpolation conditions

$$L_i(R_n) = w_i \qquad \text{for } i = 0,...,n,$$

that is

$$a_0\, L_i(x_0^*) + ... + a_n L_i(x_0^*) = w_i \qquad i = 0,...n.$$

Since $L_i(x_j^*) = 0$ for $i < j$, this system of equations reduces to a triangular one

$$a_0 L_i(x_0^*) + ... + a_i L_i(x_i^*) = w_i \qquad i = 0,...,n.$$

Thus, the a_i's are independent of the value of n which means that we have

$$R_n = R_{n-1} + a_n x_n^* \qquad\qquad n = 0,1,...$$

with

$$R_{-1} = 0.$$

Applying L_n we obtain

$$L_n(R_n) = L_n(R_{n-1}) + a_n L_n(x_n^*) = w_n$$

that is

$$a_n = \frac{w_n - L_n(R_{n-1})}{L_n(x_n^*)}$$

which is exactly the scheme given in [21].

From the determinantal expressions of R_{n-1} and x_n^* given above we immediately see that

$$a_n = \begin{vmatrix} w_n & L_n(x_0) & ... & L_n(x_{n-1}) \\ w_0 & L_0(x_0) & ... & L_0(x_{n-1}) \\ ... & ... & ... & ... \\ w_{n-1} & L_{n-1}(x_0) & ... & L_{n-1}(x_{n-1}) \end{vmatrix} \Big/ G_{n+1}.$$

Using the Schur complement technique as explained in [28], we have

$$R_n = A_n^{-1} W_n * X_n$$

where

$$A_n = \begin{pmatrix} L_0(x_0) & ... & L_0(x_n) \\ ... & ... & ... \\ L_n(x_0) & ... & L_n(x_n) \end{pmatrix}$$

$$W_n = (w_0,...,w_n)^T$$

$$X_n = (x_0,...,x_n)^T.$$

Thus $A_n^{-1} W_n$ is a vector d with components $d_0,...,d_n$ and the notation d $* X_n$ denotes the linear combination $d_0 x_0 + ... + d_n x_n$.

Since the preceding formula for R_n generalizes Newton's, then a_n is a generalization of the classical divided differences. A recursive scheme for their computation will be given later (see section 4.1).

The interpolation conditions can be written as

$$L_i(R_n - f) = 0 \qquad\qquad i = 0,...,n.$$

Moreover, let I_n be the linear mapping on E defined by

$$I_n f = R_n.$$

We have $I_n R_n = R_n$ and thus $I_n^2 = I_n$ which shows that I_n is a projection on E_n and that R_n is the truncated formal expansion of f corresponding to the biorthogonal family $\{L_i^*, x_j^*\}$ (a generalization of the Fourier expansion in a Hilbert space). The convergence of such expansions has been the subject of vaste investigations which shall not be discussed here (see, for example, [177]).

Let us only mention a generalization of a well known minimization property showing that, in a Hilbert space, the truncated Fourier series is the solution of the best approximation problem

Theorem 6 : $\forall f \in E, \forall i \geq 0$

$$|L_i^*(f-R_n)| \leq |L_i^*(f - \sum_{j=o}^{n} \alpha_j\, x_j^*)|$$

for all possible choices of $\alpha_0,...,\alpha_n$.

Proof : For $0 \leq i \leq n$ we have

$$L_i^*(R_n) = \sum_{j=o}^{n} L_j^*(f)\, L_i^*(x_j^*) = L_i^*(f)$$

and thus the left hand side of the inequality is zero. In the right hand side we have

$$L_i^*(\sum_{j=0}^{n} \alpha_j \ x_j^*) = \alpha_i$$

which shows that the best possible choice for α_i is $\alpha_i = L_i^*(f)$. For $i > n$ the inequality reduces to the equality $|L_i^*(f)| = |L_i^*(f)|$ since $L_i^*(x_j) = 0$ for $i > n \geq j$. ◆

Let $p = \alpha_0 x_0^* + ... + \alpha_n x_n^*$. Then $L_i^*(p) = \alpha_i$ for $i = 0,...,k$ and $= 0$ for $i > k$.

Up to now we have always been dealing with the interpolation problem in E_n. Similarly the dual interpolation problem in E_n^* can be studied. It consists in finding $M_n \in E_n^* = $ Span $(L_0,...,L_n)$ such that

$$M_n(x_i) = v_i \qquad \text{for } i = 0,...,n$$

for arbitrary values of $v_0,...,v_n$ not all zero.

This dual interpolation problem has a unique solution under the same assumptions as above. As for R_n, M_n will be more easily constructed via the Newton's basis that is M_n will be written as

$$M_n = b_0 L_0^* + ... + b_n L_n^*.$$

Thus the interpolation conditions are
$$b_0 L_0^*(x_i) + ... + b_n L_n^*(x_i) = v_i \qquad \text{for } i = 0,...,n.$$

Since $L_j^*(x_i) = 0$ for $i = 0,...,j-1$ and $L_i^*(x_i) = 1$, the preceding system reduces to a triangular one (with a unit diagonal)

$$b_0 L_0^*(x_i) + ... + b_i L_i^*(x_i) = 0 \qquad i = 0,...,n.$$

Thus the b_i's are independent of n which means that

$$M_n = M_{n-1} + b_n L_n^* \qquad n = 0,1,...$$

with

$$M_{-1} = 0.$$

Moreover

$$M_n(x_n) = M_{n-1}(x_n) + b_n L_n^*(x_n) = v_n$$

and thus

$$b_n = v_n - M_{n-1}(x_n).$$

We also have

$$b_n = M_n(x_n^*)$$

and

$$b_n = \begin{vmatrix} v_n & L_0(x_n) & ... & L_{n-1}(x_n) \\ v_0 & L_0(x_0) & ... & L_{n-1}(x_0) \\ ... & ... & ... & ... \\ v_{n-1} & L_0(x_{n-1}) & ... & L_{n-1}(x_{n-1}) \end{vmatrix} / G_n.$$

M_n can also be expressed as a ratio of determinants or via the Schur complement

$$M_n = - \begin{vmatrix} 0 & L_0 & ... & L_n \\ v_0 & L_0(x_0) & ... & L_n(x_0) \\ ... & ... & ... & ... \\ v_n & L_0(x_n) & ... & L_n(x_n) \end{vmatrix} / G_{n+1}$$

$$= (A_n^T)^{-1} V_n * Z_n$$

with A_n as above, $V_n = (v_0,...,v_n)^T$ and $Z_n = (L_0,...,L_n)^T$. The b_n's are generalized divided differences in the dual space and we shall give a recursive scheme for their computation in section 4.1.

If we set $v_i = L(x_i)$ then $b_n = L(x_n^*)$.

The dual interpolation conditions can be written as

$$(M_n-L)(x_i) = 0 \qquad\qquad i = 0,...,n.$$

Moreover, let J_n be the linear mapping on E^* defined by

$$J_n L = M_n.$$

We have $J_n M_n = M_n$ and thus $J_n^2 = J_n$ which shows that J_n is a projection on E_n^*. The connection between I_n and J_n will be studied in section 3.4.

Of course $M_n = \sum_{i=0}^{n} L(x_i^*)L_i^*$ can be considered as the truncated formal expansion of L corresponding to the biorthogonal family $\{L_i^*, x_j^*\}$. For such expansions we have a result similar to theorem 6

Theorem 7 : $\forall L \in E^*, \forall i \geq 0$

$$| (L-M_n)(x_i^*) | \leq | (L - \sum_{j=0}^{n} \beta_j L_j^*)\, (x_i^*)|$$

for all possible choices of $\beta_0,...,\beta_n$.

Let $e = \beta_0 L_0^* + ... + \beta_n L_n^*$. Then $e\,(x_i^*) = \beta_i$ for $i = 0,...,k$ and $= 0$ for $i > k$.

On interpolation and projection see also [171].

3.3 - Kernel.

By analogy with orthogonal polynomials and with the Christoffel-Darboux identity let us define the kernel $K_n(L,f)$ by

$$K_n(L,f) = - \begin{vmatrix} 0 & L_0(f) & ... & L_n(f) \\ L(x_0) & L_0(x_0) & ... & L_n(x_0) \\ ... & ... & ... & ... \\ L(x_n) & L_0(x_n) & ... & L_n(x_n) \end{vmatrix} / G_{n+1}.$$

This is a bilinear form on $E^* \times E$ such that if $w_i = L_i(f)$ and $v_i = L(x_i)$ for $i = 0,...,n$ we have

$$K_n(L,\bullet) = M_n$$

$$K_n(\bullet,f) = R_n.$$

In that case we also have

$$K_n(L,f) = M_n(f) = L(R_n).$$

Thus

$$K_n(L,f) = b_0 L_0^*(f) + \ldots + b_n L_n^*(f)$$

$$= a_0 L(x_0^*) + \ldots + a_n L(x_n^*).$$

As we saw before $a_i = L_i^*(f)$ and $b_i = L(x_i^*)$ and it follows that

$$K_n(L,f) = \sum_{i=o}^{n} L(x_i^*) L_i^*(f)$$

that is

$$K_n(L,f) = K_{n-1}(L,f) + L(x_n^*) L_n^*(f) \qquad\qquad n = 0,1,\ldots$$

with

$$K_{-1}(L,f) = 0.$$

Using our previous notations we have

$$K_n(L,f) = (W_n, A_n^{-1} V_n).$$

The following properties hold

$$K_n(L_i^*, x_j^*) = \delta_{ij}$$

$$K_n(L, x_j^*) = \begin{cases} L(x_j^*) & j \le n \\ 0 & j > n \end{cases}$$

$$K_n(L_j^*, f) = \begin{cases} L_j^*(f) & j \le n \\ 0 & j > n \end{cases}$$

Let $p = \alpha_0 x_0^* + \ldots + \alpha_n x_n^*$. Then, for $i = 0,\ldots,n$, $L_i^*(p) = \alpha_i$.
Thus

$$K_n(L,p) = \sum_{i=o}^{n} \alpha_i \, L(\overset{*}{x_i}) = L(p) \qquad \forall L \in E^*.$$

Similarly let $e = \beta_0 \overset{*}{L_0} + ... + \beta_n \overset{*}{L_n}$. Then, for $i = 0,...,n$, $e(\overset{*}{x_i}) = \beta_i$ and thus

$$K_n(e,f) = \sum_{i=o}^{n} \beta_i \overset{*}{L_i}(f) = e(f) \qquad \forall f \in E.$$

The first of these properties is noting else than the classical reproducing property of K_n when E is a commutative algebra and when $L_i(f) = c(x_i \, f)$. In that case, $\forall f \in E$

$$\overset{*}{L_i}(f) = a_{i0}L_0(f) + ... + a_{ii}L_i(f)$$

$$= \frac{G_i}{G_{i+1}} \, c((b_{i0}x_0 + ... + b_{ii}x_i) \, f)$$

$$= \frac{G_i}{G_{i+1}} \, c(\overset{*}{x_i} \, f).$$

If $p = \alpha_0 \overset{*}{x_0} + ... + \alpha_n \overset{*}{x_n}$ then $\overset{*}{L_i}(p) = \alpha_i = \frac{G_i}{G_{i+1}} \, c(\overset{*}{x_i} \, p)$

and we obtain

$$K_n(L,p) = c(p \sum_{i=o}^{n} \frac{G_i}{G_{i+1}} \, \overset{*}{x_i} \, \overset{*}{L}(\overset{*}{x_i})) = L(p).$$

In the case $x_i = x^i$ and if L is defined by $L(p) = p(t)$ where p is an arbitrary polynomial, then the preceding relation is exactly the reproducing property of K_n. Thus we have extended this property to a more general setting.

In the case of a commutative algebra and $L_i(f) = c(x_i f)$ we also have $\forall p, \, f \in E$

$$c(p \, K_n(\bullet,f)) = c(f \, K_n(\bullet,f)).$$

Let us now give a kind of generalization of the Christoffel-Darboux identity. It also generalizes the formula given by Iserles and Nørsett [111] for biorthogonal polynomials.

We set $\forall x, y \in E$

$$H_{n+1}(x,y) = \begin{vmatrix} L_0(x) & \dots & L_{n+1}(x) \\ L_0(y) & \dots & L_{n+1}(y) \\ L_0(x_0) & \dots & L_{n+1}(x_0) \\ \dots & \dots & \dots \\ L_0(x_{n-1}) & \dots & L_{n+1}(x_{n-1}) \end{vmatrix}.$$

Applying Schweins' determinantal formula we obtain

$$H_{n+1}(x,y) = G_{n+2}[L_{n+1}^*(y) \ L_n^*(x) - L_{n+1}^*(x) \ L_n^*(y)].$$

Similarly if we set $\forall L, e \in E^*$

$$F_{n+1}(e,L) = \begin{vmatrix} e(x_0) & \dots & e(x_{n+1}) \\ L(x_0) & \dots & L(x_{n+1}) \\ L_0(x_0) & \dots & L_0(x_{n+1}) \\ \dots & \dots & \dots \\ L_{n-1}(x_0) & \dots & L_{n-1}(x_{n+1}) \end{vmatrix}$$

and if we apply Schweins' identity we obtain

$$F_{n+1}(e,L) = G_n[e(x_{n+1}^*) \ e(x_n^*) - e(x_{n+1}^*) \ L(x_n^*)].$$

These two formulae correspond to the second Christoffel-Darboux-type formula given in the above mentioned paper of Iserles and Nørsett [111]. Similarly if, in H_{n+1} and F_{n+1} the last columns are put in the first position and if the first rows are placed as last ones, formulae corresponding to the first Christoffel-Darboux-type formula of [111] are obtained by application of Sylvester's determinantal formula, (on these two determinantal formulae, see appendix 3).

3.4 - The interpolation operator.

In section 3.2 we already defined the linear mappings I_n and J_n. Let

$$I_n : E \to E_n \text{ such that } I_n(f) = R_n$$

$$J_n : E^* \to E_n^* \text{ such that } J_n(L) = M_n.$$

From the preceding determinantal formulae we have

$$I_n(\bullet) = - \begin{vmatrix} 0 & x_0 & ... & x_n \\ L_0(\bullet) & L_0(x_0) & ... & L_0(x_n) \\ ... & ... & ... & ... \\ L_n(\bullet) & L_n(x_0) & ... & L_n(x_n) \end{vmatrix} / G_{n+1}$$

$$J_n(\bullet) = - \begin{vmatrix} 0 & L_0 & ... & L_n \\ <\bullet,x_0> & L_0(x_0) & ... & L_n(x_0) \\ ... & ... & ... & ... \\ <\bullet,x_n> & L_0(x_n) & ... & L_n(x_n) \end{vmatrix} / G_{n+1}$$

where $<\bullet,\bullet>$ denotes the duality between E^* and E, that is $\forall L \in E^*$, $\forall f \in E$, $<L,f> = L(f)$. In the sequel we shall make use simultaneously of these two notations according to the circumstances.

$\forall (L,f) \in E^* \times E$ we have

$$<L, I_n(f)> = <J_n(L), f> = K_n(L,f).$$

Thus, by definition, J_n is the dual operator of I_n, that is

$$J_n = I_n^*.$$

We have

$$f-I_n(f) = \begin{vmatrix} f & x_0 & ... & x_n \\ L_0(f) & L_0(x_0) & ... & L_0(x_n) \\ ... & ... & ... & ... \\ L_n(f) & L_n(x_0) & ... & L_n(x_n) \end{vmatrix} / G_{n+1}$$

and thus

$$<L_i, f - I_n(f)> = 0 \qquad\qquad i = 0,...,n$$

which is equivalent to the interpolation conditions

$$<L_i, f> = <L_i, I_n(f)> = <L_i, R_n> \qquad\qquad i = 0,...,n.$$

Similarly we have

$$L-I_n^*(L) = \begin{vmatrix} L & L_0 & ... & L_n \\ L(x_0) & L_0(x_0) & ... & L_n(x_0) \\ ... & ... & ... & ... \\ L(x_n) & L_0(x_n) & ... & L_n(x_n) \end{vmatrix} / G_{n+1}.$$

Thus

$$<L - I_n^*(L), x_i> = 0 \qquad\qquad i = 0,...,n$$

which is equivalent to the dual interpolation conditions

$$<L, x_i> = <I_n^*(L), x_i> = <M_n, x_i> \qquad\qquad i = 0,...,n.$$

Of course, since I_n and I_n^* are projections on E_n and E_n^* respectively, we have for $i = 0,...,n$

$$x_i = I_n(x_i)$$
$$L_i = I_n^*(L_i).$$

As we previously saw, $\forall f \in E$, the series $\sum_{i=0}^{\infty} L_i^*(f)x_i^*$ is called the formal expansion of f corresponding to the biorthogonal family $\{L_i^*, x_j^*\}$ and we write [177] :

$$f \sim \sum_{i=0}^{\infty} L_i^*(f) x_i^*.$$

We have

$$I_n(f) = \sum_{i=0}^{n} L_i^*(f) x_i^*.$$

Replacing f by its approximation $I_n(f)$ is known as Galerkin's method. We have

$$f - I_n(f) \sim \sum_{i=n+1}^{\infty} L_i^*(f) x_i^*.$$

Thus

$$x_{n+1} - I_n(x_{n+1}) = x_{n+1}^*, \quad L_i(x_{n+1} - I_n(x_{n+1})) = 0 \qquad i = 0,...,n$$

and

$$x_k^* - I_n(x_k^*) = \begin{cases} 0 & k \leq n \\ x_k^* & k > n. \end{cases}$$

Of course, similar results hold in E^*.

As we saw in the introduction the direct problem consists in computing the numerical value of L(f). This is the case, for example, in numerical quadratures. An approximate value of L(f) can be obtained by two methods :

- replace f by I_n (f) and compute $<L, I_n(f)>$
- replace L by I_n^* (L) and compute $<I_n^*(L), f>$.

By definition of I_n^*, these two methods lead to the same approximate value of L(f) that is $<L, I_n(f)>$. This is exactly the procedure followed to obtain interpolatory quadrature formulae such as Newton-Cotes or Gaussian quadrature rules. This is also the case in Padé-type approximation as we shall see now.

Let c be the linear functional on P defined by

$$c(x^i) = c_i \qquad\qquad i = 0,1,...$$

and let us consider the formal power series

$$f(t) = \sum_{i=0}^{\infty} c_i t^i.$$

Then

$$f(t) = <c, (1-xt)^{-1}>.$$

Let v_n be an arbitrary polynomial of degree n and let R_n be the Hermite interpolation polynomial of $(1-xt)^{-1}$ at the zeros of v_n.

$<c, R_n>$ is a rational function with a numerator of degree n-1 in t and a denominator of degree n. Its series expansion in ascending powers of t agrees with that of f up to the degree n-1 that is

$$f(t) - <c, R_n> = O(t^n).$$

$<c, R_n>$ is called a Padé-type approximant of f and is denoted by

$$(n-1/n)_f(t).$$

If v_n is the polynomial of degree n belonging to the family of formal orthogonal polynomials with respect to c (that is satisfying $c(x^i v_n(x)) = 0$ for $i = 0,...,n-1$) then

$$f(t) - <c, R_n> = O(t^{2n})$$

and in that case $<c, R_n>$ is called a Padé approximant of f and is denoted by $[n-1/n]_f(t)$. Thus Padé approximants appear as formal Gaussian quadrature methods for the function $(1-xt)^{-1}$. This point of view was developed in [17] (see also [38], which is more recent). If I_n is defined as

$$R_n = I_n((1-xt)^{-1})$$

then

$$<c, R_n> = <c, I_n((1-xt)^{-1})>$$

The linear functional $d = I_n^*(c)$ is studied in details in appendix 2.

A well known method for estimating the error $L(f) - L(R_n)$ in Gaussian quadrature methods is Kronrod's procedure [118].
Since Padé approximants are formal Gaussian methods, Kronrod's procedure can be extended to Padé approximants to estimate their error [27]. It can now be extended to our general setting.

Let $L(R_n)$ and $L(R_{n+m})$ be two approximations of $L(f)$. Then

$$\frac{L(R_{n+m})-L(R_n)}{L(f)-L(R_n)} = 1 - \frac{L(R_{n+m})-L(f)}{L(R_n)-L(f)}.$$

If $|L(R_{n+m}) - L(f)| << |L(R_n) - L(f)|$ (which is the case if (R_n) converges weakly to f) then $L(R_{n+m}) - L(R_n)$ is a good approximation of the error $L(f) - L(R_n)$.

Since

$$R_{n+i} = R_{n+i-1} + L_{n+i}^*(f) \, x_{n+i}^*$$

then

$$L(R_{n+m}) - L(R_n) = \sum_{i=1}^{m} L_{n+i}^*(f) \, L(x_{n+i}^*)$$

which is an extension of the diagonal expansion of the error used by Belantari [8] for estimating the error in Padé approximation.

3.5 - The method of moments.

This method, studied by Vorobyev [186] in a Hilbert space, is a particular case of Galerkin's method. We shall now extend it to an arbitrary vector space E and its dual E^*.

The method of moments consists in constructing a linear operator A_n on E_{n-1} such that

$$x_1 = A_n x_0$$
$$x_2 = A_n x_1$$
$$\text{-----------}$$
$$x_{n-1} = A_n x_{n-2}$$
$$I_{n-1}(x_n) = A_n x_{n-1}$$

or

$$x_k = A_n^k x_0 \qquad\qquad k = 0,...,n-1$$
$$I_{n-1}(x_n) = A_n^n x_0.$$

Let $x \in E_{n-1}$. Then

$$x = c_0 x_0 + ... + c_{n-1} x_{n-1}.$$

Thus

$$A_n x = c_0 A_n x_0 + ... + c_{n-2} A_n x_{n-2} + c_{n-1} A_n x_{n-1}$$

$$= c_0 x_1 + .. + c_{n-2} x_{n-1} + c_{n-1} I_{n-1}(x_n) \in E_{n-1}.$$

Since $I_{n-1}(x_n) \in E_{n-1}$, we can find $\alpha_0,...,\alpha_{n-1}$ such that

$$I_{n-1}(x_n) = -\alpha_0 x_0 - ... - \alpha_{n-1} x_{n-1}$$

that is

$$\alpha_0 x_0 + ... + \alpha_{n-1} x_{n-1} + I_{n-1}(x_n) = (\alpha_0 I + \alpha_1 A_n + ... + \alpha_{n-1} A_n^{n-1} + A_n^n) x_0 = 0$$

where I is the identity mapping in E.

We have, as we saw in the previous section

$$L_i(x_n - I_{n-1}(x_n)) = 0 \qquad\qquad i = 0,...,n-1$$

that is

$$\alpha_0 L_i(x_0) + ... + \alpha_{n-1} L_i(x_{n-1}) + L_i(x_n) = 0 \qquad\qquad \text{for } i = 0,...,n-1.$$

This system has a unique solution since its determinant G_n is different from zero.

Let us set

$$P_n(t) = \alpha_0 + \alpha_1 t + ... + \alpha_{n-1} t^{n-1} + t^n.$$

We have

$$P_n(A_n)x_0 = \alpha_0 x_0 + ... + \alpha_{n-1} x_{n-1} + I_{n-1}(x_n) = 0.$$

Now let λ be an eigenvalue of A_n and let u be the corresponding eigenelement. $u \in E_{n-1}$ and thus

$$u = a_0 x_0 + ... + a_{n-1} x_{n-1}.$$

Then

$$A_n u = a_0 A_n x_0 + ... + a_{n-1} A_n x_{n-1} = \lambda(a_0 x_0 + ... + a_{n-1} x_{n-1})$$

$$= a_0 x_1 + ... + a_{n-2} x_{n-1} + a_{n-1} I_{n-1}(x_n)$$

$$= a_0 x_1 + ... + a_{n-2} x_{n-1} + a_{n-1}(-\alpha_0 x_0 - ... - \alpha_{n-1} x_{n-1}).$$

Thus

$$- \alpha_0 a_{n-1} x_0 + (a_0 - \alpha_1 a_{n-1}) x_1 + ... + (a_{n-2} - \alpha_{n-1} a_{n-1}) x_{n-1} =$$

$$a_0 \lambda x_0 + ... + a_{n-1} \lambda x_{n-1}.$$

Since $x_0,...,x_{n-1}$ are independent in E_{n-1} we must have

$$- \alpha_0 a_{n-1} = a_0 \lambda$$

$$a_i - \alpha_{i+1} a_{n-1} = a_{i+1} \lambda \qquad i = 0,...,n-2$$

that is, in matricial form

$$
\begin{pmatrix}
-\lambda & 0 & 0 & ... & 0 & 0 & -\alpha_0 \\
1 & -\lambda & 0 & ... & 0 & 0 & -\alpha_1 \\
- & - & - & - & - & - & - \\
0 & 0 & 0 & ... & 1 & -\lambda & -\alpha_{n-2} \\
0 & 0 & 0 & ... & ... & 1 & (-\alpha_{n-1}-\lambda)
\end{pmatrix}
\begin{pmatrix}
a_0 \\
a_1 \\
\vdots \\
\\
a_{n-2} \\
a_{n-1}
\end{pmatrix}
= 0.
$$

In order for this system to have a non trivial solution, its determinant must be zero, that is

$$P_n(\lambda) = 0$$

which shows that P_n is the characteristic polynomial of A_n. Moreover, $a_{n-1} \neq 0$ since, otherwise, all the a_i's would be zero. Since an eigenelement is determined apart from a multiplying factor, we can choose $a_{n-1} = 1$ and we have

$$a_{n-2} = \alpha_{n-1} + \lambda$$

$$a_i = \alpha_{i+1} + a_{i+1}\lambda \qquad i = n-3,...,0.$$

All the other results concerning the method of moments also follow and, in particular, those concerning the solution of operator equations (that is the inverse problem of the introduction) [17, pp. 76-77].

We consider the equation $f = A_n f + b$ in E_{n-1} (that is f, b$\in$ E_{n-1}).
Let P and Q be two polynomials related by

$$1 - P(t) = (1-t)\, Q(t).$$

Then the degree of Q is one less than the degree of P, $P(1) = 1$ and we have

$$f = P(A_n)f + Q(A_n)b.$$

If we choose P as $P(t) = P_n(t)/P_n(1)$ where P_n is the polynomial defined above (that is the characteristic polynomial of A_n) then $P_n(1) \neq 0$ since $I - A_n$ is invertible and we have

$$f = Q(A_n)\, b.$$

If we set

$$P(t) = a_0 + a_1 t + ... + a_n t^n$$

$$Q(t) = b_0 + b_1 t + ... + b_{n-1} t^{n-1}$$

then $a_i = \alpha_i / \sum_{i=0}^{n} \alpha_i$ with $\alpha_n = 1$ and $b_i = \sum_{j=i+1}^{n} a_j$ for $i = 0,...,n-1$.

Another possible approach is to write

$$b = c_0 x_0 + ... + c_{n-1} x_{n-1}$$

$$f = d_0 x_0 + ... + d_{n-1} x_{n-1}$$

where the c_i's are solution of the system

$$L_i(b) = c_0 L_i(x_0) + ... + c_{n-1} L_i(x_{n-1}) \qquad i = 0,...,n-1.$$

Thus

$$d_0 x_0 + ... + d_{n-1} x_{n-1}$$

$$= d_0 A_n x_0 + ... + d_{n-2} A_n x_{n-2} + d_{n-1} A_n x_{n-1} + c_0 x_0 + ... + c_{n-1} x_{n-1}$$

$$= d_0 x_1 + ... + d_{n-2} x_{n-1} + d_{n-1} I_{n-1}(x_n) + c_0 x_0 + ... + c_{n-1} x_{n-1}.$$

Replacing $I_{n-1}(x_n)$ by $-\alpha_0 x_0 - ... - \alpha_{n-1} x_{n-1}$ and equating the coefficients of $x_0,...,x_{n-1}$ we obtain

$$d_0 = -d_{n-1} \alpha_0 + c_0$$

$$d_i = d_{i-1} - d_{n-1}\alpha_i + c_i \qquad i = 1,...,n-1.$$

Summing up these relations we get

$$d_0 + ... + d_{n-1} = d_0 + ... + d_{n-2} - d_{n-1}(\alpha_0 + ... + \alpha_{n-1}) + c_0 + ... + c_{n-1}$$

and thus

$$d_{n-1} = (c_0 + ... + c_{n-1})/(\alpha_0 + ... + \alpha_{n-1} + 1)$$

and then $d_0, d_1,...,d_{n-2}$ are directly obtained from the preceding relations.

Now let us solve $A_n f = b$ in E_{n-1}. We choose P and Q related by

$$1 - P(t) = t\, Q(t).$$

The degree of Q is one less than that of P and $P(0) = 1$. If we choose $P(t) = P_n(t)/P_n(0)$, which is possible since A_n is invertible and thus $P_n(0) \neq 0$ then

$$f = Q(A_n)b.$$

The coefficients a_i of P are $a_i = \alpha_i/\alpha_0$ and those of Q are given by $b_i = -a_{i+1}$ for $i = 0,...,n-1$.

Writing again b and f as above, the second approach leads to

$$d_0 A_n x_0 + ... + d_{n-2} A_n x_{n-2} + d_{n-1} A_n x_{n-1} = c_0 x_0 + ... + c_{n-1} x_{n-1}$$

$$= d_0 x_1 + ... + d_{n-2} x_{n-1} + d_{n-1} I_{n-1}(x_n) = c_0 x_0 + ... + c_{n-1} x_{n-1}$$

and thus we obtain

$$d_{n-1} = -c_0/\alpha_0$$

and then

$$d_i = c_{i+1} + d_{n-1}\alpha_{i+1} \qquad\qquad i = 0,...,n-2.$$

Let A be an operator in E. x_0 being given we assume that the x_i's are formed by

$$x_{i+1} = Ax_i \qquad\qquad i = 0,1,...$$

and that $x_0,...,x_n$ are linearly independent. The operator A_n constructed by the preceding generalization of the method of moments is such that

$$A_n = I_n A I_n$$

which means that $\forall f \in E$, $A_n f = I_n A I_n f$.

We also have

$$P_n(t) = \begin{vmatrix} L_0(x_0) & ... & L_0(x_{n-1}) & L_0(x_n) \\ ... & ... & ... & ... \\ L_{n-1}(x_0) & ... & L_{n-1}(x_{n-1}) & L_{n-1}(x_n) \\ 1 & ... & t^{n-1} & t^n \end{vmatrix} \Big/ G_n.$$

A generalization of the method of moments can also be defined in E^*. We want to construct a linear operator B_n on E_{n-1}^* such that

$$L_1 = B_n L_0$$
$$L_2 = B_n L_1$$
$$\text{----------}$$
$$L_{n-1} = B_n L_{n-2}$$
$$I_{n-1}^*(L_n) = B_n L_{n-1}$$

or

$$L_k = B_n^k L_o \qquad\qquad k = 0,...,n-1$$

$$I_{n-1}^* (L_n) = B_n^n L_o.$$

Let $L \in E_{n-1}^*$. Then

$$L = d_0 L_0 + ... + d_{n-1} L_{n-1}.$$

Thus

$$B_n L = d_0 B_n L_0 + ... + d_{n-2} B_n L_{n-2} + d_{n-1} B_n L_{n-1}$$

$$= d_0 L_1 + ... + d_{n-2} L_{n-1} + d_{n-1} I_{n-1}^* (L_n).$$

Since $I_{n-1}^* (L_n) \in E_{n-1}^*$, we can find $\beta_0,...,\beta_{n-1}$ such that

$$I_{n-1}^* (L_n) = - \beta_0 L_0 - .... - \beta_{n-1} L_{n-1}$$

that is

$$\beta_0 L_0 + ... + \beta_{n-1} L_{n-1} + I_{n-1}^*(L_n) = (\beta_0 I^* + \beta_1 B_n + ... + \beta_{n-1} B_n^{n-1} + B_n^n) L_0 = 0$$

where I^* is the identity operator in E^*.

We have

$$(L_n - I_{n-1}^*(L_n)) (x_i) = 0 \qquad\qquad \text{for } i = 0,...,n-1$$

that is

$$\beta_0 L_0(x_i) + ... + \beta_{n-1} L_{n-1}(x_i) + L_n(x_i) = 0 \qquad i = 0,...,n-1.$$

This system has a unique solution since its determinant G_n is different from zero.

Let us set

$$Q_n(t) = \beta_0 + \beta_1 t + ... + \beta_{n-1} t^{n-1} + t^n.$$

Let μ be an eigenvalue of B_n and let v be the corresponding eigenelement. $v \in E_{n-1}^*$ and thus

$$v = b_0 L_0 + ... + b_{n-1} L_{n-1}.$$

Then

$$B_n v = b_0 B_n L_0 + ... + b_{n-1} B_n L_{n-1} = \mu (b_0 L_0 + ... + b_{n-1} L_{n-1})$$

$$= b_0 L_1 + ... + b_{n-2} L_{n-1} + b_{n-1} I_{n-1}^* (L_n)$$

$$= b_0 L_1 + ... + b_{n-2} L_{n-1} + b_{n-1} (-\beta_0 L_0 - ... - \beta_{n-1} L_{n-1}).$$

Thus

$$-\beta_0 b_{n-1} L_0 + (b_0 - \beta_1 b_{n-1}) L_1 + ... + (b_{n-2} - -\beta_{n-1} b_{n-1}) L_{n-1} =$$

$$b_0 \mu L_0 + ... + b_{n-1} \mu L_{n-1}.$$

Since $L_0,...,L_{n-1}$ are independent in E_{n-1}^* we must have

$$-\beta_0 b_{n-1} = b_0 \mu$$

$$b_i - \beta_{i+1} b_{n-1} = b_{i+1} \mu \qquad\qquad i = 0,...,n-2.$$

This system has a non trivial solution if and only if its determinant is zero, that is

$$Q_n(\mu) = 0$$

which shows that Q_n is the characteristic polynomial of B_n

$$Q_n(t) = \begin{vmatrix} L_0(x_0) & ... & L_n(x_0) \\ ... & ... & ... \\ L_0(x_{n-1}) & ... & L_n(x_{n-1}) \\ 1 & ... & t^n \end{vmatrix} / G_n.$$

If B is an operator in E^* such that $L_{i+1} = B L_i$ for $i = 0,1,...,$ then the operator B_n constructed by the method of moments is

$$B_n = I_n^* B I_n^*.$$

The operators A_n and B_n are approximations of the operators A and B respectively thus solving the identification problem mentioned in the introduction.

3.6 - Lanczos' method.

In a Hilbert space it is well known that the method of moments gives rise to Lanczos' method and then to the conjugate and bi-conjugate gradient methods, see [17 , pp. 79-91, 186-189]. The generalization of Lanczos' method to our setting will be studied in this section and that of the bi-conjugate gradient method in the next one.

Let $x_0 \in E$ and $L_0 \in E^*$ be given and let A be a linear operator on E. We assume that, for $i = 0,1,...$

$$x_{i+1} = Ax_i$$
$$L_{i+1} = A^*L_i$$

where A^* is the dual of A. E is also assumed to be reflexive so that $A^{**} = A$.

We have

$$\langle L_i, x_j \rangle = \langle A^{*i}L_0, A^j x_0 \rangle = \langle L_0, A^{i+j}x_0 \rangle$$

$$= \langle A^{*j}L_0, A^i x_0 \rangle = \langle L_j, A^i x_0 \rangle = \langle L_j, x_i \rangle = \langle L_k, x_m \rangle$$

$$= c_{i+j}$$

if $m+k = i+j$.

Let P_n and Q_n be the polynomials obtained by the method of moments applied to $(x_0,...,x_n)$ and $(L_0,...,L_n)$ respectively. These polynomials are identical since they are given by the linear systems

$$\alpha_0 L_i(x_0) + ... + \alpha_{n-1} L_i(x_{n-1}) + L_i(x_n) = 0 \qquad i = 0,...,n-1$$
$$\beta_0 L_0(x_i) + ... + \beta_{n-1} L_{n-1}(x_i) + L_n(x_i) = 0 \qquad i = 0,...,n-1.$$

Let c be the linear functional on P defined by

$$c(x^k) = c_k = \langle L_i, x_j \rangle \quad \text{with } i+j = k.$$

Then the preceding system can be writen as

$$\alpha_0 c_i + ... + \alpha_{n-1} c_{i+n-1} + c_{i+n} = 0 \qquad i = 0,...,n-1$$

that is

$$c(x^i(P_n(x))) = 0 \qquad\qquad \text{for } i = 0,...,n-1$$

which shows that $\{P_n\}$ is the family of formal orthogonal polynomials with respect to c. Thus $\{P_n\}$ satisfies the usual three-terms recurrence relationship

$$P_{n+1}(x) = (x + B_{n+1})P_n(x) - C_{n+1}P_{n-1}(x) \qquad\qquad n = 0,1,...$$

with

$$B_{n+1} = - c(xP_n^2(x))/c(P_n^2(x)) \qquad\qquad C_{n+1} = c(P_n^2(x))/c(P_{n-1}^2(x)).$$

Let us express these constants. We have

$$c(P_n^2(x)) = <L_0, P_n^2(A)x_0> = <P_n(A^*)L_0, P_n(A)x_0>$$

We set

$$\hat{x}_n = P_n(A)x_0$$
$$\hat{L}_n = P_n(A^*)L_0.$$

Thus

$$c(P_n^2(x)) = <\hat{L}_n, \hat{x}_n> \quad \text{and} \quad c(xP_n^2(x)) = <\hat{L}_n, A\hat{x}_n>,$$

and it follows that

$$B_{n+1} = - <\hat{L}_n A\hat{x}_n>/<\hat{L}_n, \hat{x}_n>$$
$$C_{n+1} = <\hat{L}_n, \hat{x}_n> /<\hat{L}_{n-1}, \hat{x}_{n-1}>.$$

We have

$$P_{n+1}(A) = (A + B_{n+1}) P_n(A) - C_{n+1}P_{n-1}(A)$$

and an analogous relation for $P_{n+1}(A^*)$. Applying to x_0 and L_0 respectively, we obtain for $n = 0,1,...$

$$\hat{x}_{n+1} = (A+B_{n+1}) \hat{x}_n - C_{n+1} \hat{x}_{n-1}$$

$$\hat{L}_{n+1} = (A^* + B_{n+1}) \hat{L}_n - C_{n+1}\hat{L}_{n-1}$$

with

$$\hat{x}_{-1} = 0 \in E \qquad\qquad \hat{x}_0 = x_0$$
$$\hat{L}_{-1} = 0 \in E^* \qquad\qquad \hat{L}_0 = L_0.$$

The orthogonality relation $c(P_k(x) \, P_n(x)) = 0$ for $k \neq n$ is equivalent to

$$\langle \hat{L}_n, \hat{x}_n \rangle = 0.$$

Moreover

$$\hat{x}_n = P_n(A)x_0 = \begin{vmatrix} L_0(x_0) & \dots & L_0(x_n) \\ \dots & \dots & \dots \\ L_{n-1}(x_0) & \dots & L_{n-1}(x_n) \\ x_0 & \dots & x_n \end{vmatrix} \Big/ G_n = x_n^*$$

$$\hat{L}_n = P_n(A^*)L_0 = \begin{vmatrix} L_0(x_0) & \dots & L_n(x_0) \\ \dots & \dots & \dots \\ L_0(x_{n-1}) & \dots & L_n(x_{n-1}) \\ L_0 & \dots & L_n \end{vmatrix} \Big/ G_n = \frac{G_{n+1}}{G_n} L_n^*.$$

Thus Lanczos' method have been generalized in a reflexive vector space and it constructs, in a particular case, the biorthogonal family $\{L_i^*, x_j^*\}$. Moreover $A_n = I_n A I_n$ and $A_n^* = I_n^* A^* I_n^*$.

3.7 - The bi-conjugate gradient method.

We consider the inverse problem

$$Au = x_0.$$

Let $A_n = I_n A I_n$ be obtained by the method of moments and let u_n be the solution of

$$A_n u_n = x_0.$$

Let P and G be two polynomials related by

$$1 - P(t) = t \, G(t).$$

Then

$$u_n = P(A_n) \, u_n + G(A_n)x_0$$

since

$$(I-P(A_n)) \, u_n = G(A_n) \, x_o = G(A_n) \, A_n u_n.$$

The relation between P and G shows that we must have $P(0) = 1$. We shall now make the choice

$$P(t) = P_n(t)/P_n(0)$$

where P_n is the polynomial given by the method of Lanczos. The corresponding polynomial G will now be called G_n. $P_n(0) \neq 0$ since A_n is invertible and $P(0) = 1$. Then

$$u_n = G_n \, (A_n) \, x_o$$

which shows that $G_n \, (A_n) = A_n^{-1}$ where G_n is such that

$$1 - P_n(t)/P_n(0) = t \, G_n(t).$$

If we write, as above

$$P_n(t) = \alpha_o + \alpha_1 t + \dots + \alpha_{n-1} t^{n-1} + t^n$$

and

$$G_n(t) = \gamma_o + \dots + \gamma_{n-1} t^{n-1}$$

then

$$\gamma_i = - \, \alpha_{i+1}/\alpha_o \qquad\qquad i = 0,\dots,n-2$$

$$\gamma_{n-1} = - \, 1/\alpha_o.$$

In the previous section we saw that the polynomials P_n satisfy a three-terms recurrence relationship. Let us replace in it, $P_n(t)$ by $P_n(0) \, [1-t \, G_n(t)]$. We obtain

$$P_{n+1}(0) \, [1-t \, G_{n+1}(t)] = (t + B_{n+1}) \, P_n(0)[I - t \, G_n(t)] \\ - C_{n+1} P_{n-1}(0)[I-t \, G_{n-1}(t)].$$

Replacing t by A and applying the corresponding operator to x_o leads to

$$P_{n+1}(0)[I-AG_{n+1}(A)]x_o = (A+B_{n+1})P_n(0)[I-AQ_n(A)]x_o \\ - C_{n+1}P_{n-1}(0)[I-AG_{n-1}(A)]x_o.$$

Since G_n has degree $n-1$ and since $x_k = A^k x_0 = A_n^k x_0$ for $k = 0,...,n-1$ then $G_n(A)x_0 = G_n(A_n)x_0 = u_n$. Thus the preceding relation becomes

$$P_{n+1}(0)[x_0 - Au_{n+1}] = (A + B_{n+1})P_n(0)[x_0 - Au_n] - C_{n+1}P_{n-1}(0)[x_0 - Au_{n-1}]$$

or, setting $r_n = Au_n - x_0$

$$P_{n+1}(0)\ r_{n+1} = (A + B_{n+1})\ P_n(0)r_n - C_{n+1}P_{n-1}(0)\ r_{n-1}.$$

Assuming that A is invertible we obtain

$$P_{n+1}(0)\ u_{n+1} = P_n(0)\ r_n + P_n(0)\ B_{n+1}u_n - C_{n+1}P_{n-1}(0)\ u_{n-1}.$$

Setting $\mu_n = -P_n(0)/P_{n+1}(0)$ this relation becomes

$$u_{n+1} = -\ \mu_n r_n - \mu_n B_{n+1}u_n - C_{n+1}\ \mu_{n-1}\ \mu_n u_{n-1}.$$

But

$$P_{n+1}(0) = B_{n+1}P_n(0) - C_{n+1}P_{n-1}(0)$$

that is

$$-\ \mu_n^{-1} = B_{n+1} + C_{n+1}\mu_{n-1}.$$

Adding and subtracting u_n we get

$$u_{n+1} = u_n - \mu_n r_n - \mu_n(B_{n+1} + \mu_n^{-1})u_n - C_{n+1}\ \mu_{n-1}\ \mu_n u_{n-1}$$

$$= u_n - \mu_n r_n + \mu_n C_{n+1}\ \mu_{n-1}u_n - C_{n+1}\ \mu_{n-1}\ \mu_n\ u_{n-1}$$

$$= u_n + \mu_n\ v_n$$

with

$$v_n = -r_n + \mu_{n-1}C_{n+1}(u_n - u_{n-1}).$$

Thus

$$u_{n+1} = u_n + \mu_n v_n$$

$$u_n = u_{n-1} + \mu_{n-1}v_{n-1}$$

and

$$v_n = -r_n + \mu_{n-1}^2 C_{n+1} v_{n-1}.$$

Let us set $\lambda_n = \mu_{n-1}^2 C_{n+1}$. We have

$$v_n = - r_n + \lambda_n v_{n-1} \qquad\qquad \text{with } v_{-1} = 0$$

$$u_{n+1} = u_n + \mu_n v_n.$$

But $\hat{x}_n = P_n(A)x_0$ and thus $r_n = -\hat{x}_n/P_n(0)$. Let w and w_n be respectively the solutions of

$$A^*w = L_0$$

$$A_n^*w_n = L_0.$$

We have

$$w_n = Q_n(A_n^*) L_0$$

$$\bar{r}_n = A^*w_n - L_0 = -P_n(A^*)L_0/P_n(0) = -\hat{L}_n/P_n(0).$$

Moreover

$$\langle \bar{r}_{n+1}, v_n \rangle = - \langle \bar{r}_{n+1}, r_n \rangle + \lambda_n \langle \bar{r}_{n+1}, v_{n-1} \rangle$$

$$= \lambda_n \langle \bar{r}_{n+1}, v_{n-1} \rangle$$

since

$$\langle \bar{r}_{n+1}, r_n \rangle = \langle \hat{L}_{n+1}, \hat{x}_n \rangle/P_n^2(0) = 0.$$

Thus

$$\langle \bar{r}_{n+1}, v_n \rangle = \lambda_n ... \lambda_0 \langle \bar{r}_{n+1}, v_{-1} \rangle = 0.$$

But

$$\bar{r}_{n+1} = \bar{r}_n + \mu_n A^* \bar{v}_n$$

$$\bar{v}_n = -\bar{r}_n + \lambda_n \bar{v}_{n-1} \quad \text{with} \quad \bar{v}_{-1} = 0 \in E^*$$

$$<\bar{r}_{n+1}, v_n> = 0 = <\bar{r}_n, v_n> + \mu_n <A^* \bar{v}_n, v_n>.$$

Thus

$$\mu_n = - <\bar{r}_n, v_n>/<\bar{v}_n, Av_n>$$

$$\lambda_n = <\bar{r}_n, r_n>/<\bar{r}_{n-1}, r_{n-1}>.$$

This method is a generalization of the bi-conjugate gradient method of Fletcher [70]. It solves simultaneously $A_n u_n = x_0$ and $A_n^* w_n = L_0$.

The determinantal formula given in [17, p. 87] is still valid both for u_n and w_n

$$u_n = - \begin{vmatrix} 0 & x_0 & ... & x_{n-1} \\ c_0 & c_1 & ... & c_n \\ ... & ... & ... & ... \\ c_{n-1} & c_n & ... & c_{2n-1} \end{vmatrix} \Big/ \begin{vmatrix} c_1 & ... & c_n \\ ... & ... & ... \\ c_n & ... & c_{2n-1} \end{vmatrix}$$

with $c_k = L_i(x_j)$, $i+j = k$.

If A can be factorized into the product of two operators then a geometrical interpretation (in terms of projection or, equivalently, interpolation) similar to that given in [17, pp. 87-89] can be obtained.

3.8 - Fredholm equation and Padé-type approximants.

We now consider the Fredholm equation

$$u = tAu + x_0$$

where t is a parameter, and the approximate equation

$$u_n = tA_n u_n + x_0,$$

where A_n is the operator obtained by the method of moments with $x_i = A^i x_0$.

The solution u can be formally written as a Neumann series

$$u = x_0 + tAx_0 + t^2A^2x_0 + ...$$

and we have

$$L_i(u) = L_i(x_0) + L_i(x_1)t + L_i(x_2)t^2 + ... = f_i(t).$$

We shall study $L_i(u_n)$. We have

$$u_n = (I-tA_n)^{-1} x_0$$

if t^{-1} is not an eigenvalue of A_n. We formally have

$$(I-tA_n)^{-1} = I + tA_n + t^2A_n^2 + ...$$

and thus

$$L_i(u_n) = L_i(x_0) + L_i(A_nx_0)t + ... + L_i(A_n^{n-1}x_0)t^{n-1} + ...$$

$$= L_i(x_0) + ... + L_i(x_{n-1})t^{n-1} + L_i(A_n^n x_0)t^n + ...$$

$$= L_i(u) + O(t^n).$$

Let us now look at $L_i(u_n)$ as a function of t and prove that it is a rational function.

Since E_{n-1} has dimension n, then $x_0, A_nx_0,...,A_n^n x_0$ are linearly dependent. Thus $\exists e_0,...,e_n$, not all zero, such that

$$\sum_{j=0}^{n} e_j A_n^j x_0 = 0.$$

Thus, $\forall k \geq 0$

$$\sum_{j=0}^{n} e_j A_n^{k+j} x_0 = 0.$$

Let us set $\qquad L_i(A_n^j x_0) = c_j^{(i,n)}.$

We have $\qquad c_j^{(i,n)} = L_i(A^j x_0) \qquad$ for $j = 0,...,n-1$.

Moreover, $\forall k \geq 0$

$$\sum_{j=0}^{n} e_j \, L_i(A_n^{k+j} x_0) = \sum_{j=0}^{n} e_j \, c_{k+j}^{(i,n)} = 0$$

and

$$L_i(u_n) = c_0^{(i,n)} + c_1^{(i,n)} t + c_2^{(i,n)} t^2 + ...$$

which shows that $L_i(u_n)$ is a rational function of t with a numerator of degree n-1 and a denominator of degree n.
Let us now find the expressions of this numerator and of this denominator.

Any zero t of this denominator makes $I-tA_n$ singular, which means that t^{-1} is an eigenvalue of A_n and thus a zero of the polynomial P_n obtained by the method of moments.
Thus the denominator of $L_i(u_n)$ is

$$\tilde{P}_n(t) = t^n \, P_n(t^{-1}).$$

We have (suppressing the upper indexes i and n for simplicity)

$$L_i(u_n) \, \tilde{P}_n(t) = c_0 + (c_1 + c_0\alpha_{n-1})t + (c_2 + c_1\alpha_{n-1} + c_0\alpha_{n-2})t^2 + ...$$

$$+ (c_{n-1} + c_{n-2}\,\alpha_{n-1} + ... + c_0\alpha_1)t^{n-1} + \sum_{j=0}^{\infty}(c_{n+j} + c_{n+j-1}\alpha_{n-1} + ... + c_j\alpha_0)t^{n+j}.$$

On the other hand

$$\frac{P_n(x) - P_n(t)}{x-t} = (\alpha_1 + \alpha_2 x + ... + \alpha_{n-1}x^{n-2} + x^{n-1})$$

$$+ (\alpha_2 + \alpha_3 x + ... + \alpha_{n-1}x^{n-3} + x^{n-2})t + ... + (\alpha_{n-1} + x)t^{n-2} + t^{n-1}.$$

Let e_i be the linear functional on P defined by

$$e_i(x^j) = L_i(x_j).$$

Thus $e_i(P_n(x)) = 0$ for $i = 0,...,n-1$ which shows that $\{P_n\}$ is a family of biorthogonal polynomials in the sense of Iserles and Nørsett (see section 3.1).

We set

$$Q_n^{(i)}(t) = e_i\left(\frac{P_n(x)-P_n(t)}{x-t}\right)$$

where e_i acts on the variable x, and

$$\tilde{Q}_n^{(i)}(t) = t^{n-1}Q_n^{(i)}(t^{-1}).$$

We have

$$e_i\left(\frac{P_n(x)-P_n(t)}{x-t}\right) = (\alpha_1 c_0 + \alpha_2 c_1 +... + \alpha_{n-1}c_{n-2} + c_{n-1})$$
$$+ (\alpha_2 c_0 + \alpha_3 c_1 +...+\alpha_{n-1}c_{n-3}+c_{n-2})t + ... + (\alpha_{n-1}c_0 + c_1)t^{n-2} + c_0 t^{n-1}$$

which shows that

$$L_i(u_n) = \tilde{Q}_n^{(i)}(t)/\tilde{P}_n(t)$$

and that

$$L_i(u_n)\,\tilde{P}_n(t) = \tilde{Q}_n^{(i)}(t)$$

since

$$c_k = L_i(A_n^k x_0) = L_i(A^k x_0) \qquad \text{for } k = 0,...,n-1.$$

Up to now, we proved that

$$L_i(u_n) = L_i(u) + O(t^n).$$

Thus $L_i(u_n)$ is the Padé-type approximant $(n-1/n)$ of f_i. Let us look more closely to this approximation property. We have

$$L_i(u)\,\tilde{P}_n(t) - \tilde{Q}_n^{(i)}(t) = \sum_{j=o}^{\infty}(c_{n+j}^{(i)} + c_{n+j-1}^{(i)}\alpha_{n-1} + ... + c_j^{(i)}\alpha_0)t^{n+j}$$

with $$c_j^{(i)} = L_i(A^j x_0) = L_i(x_j).$$

Thus

$$L_i(u) \, \tilde{P}_n(t) - \tilde{Q}_n^{(i)}(t) = \sum_{j=0}^{\infty} (L_i(x_{n+j}) + \alpha_{n-1} L_i(x_{n+j-1}) + \ldots + \alpha_0 L_i(x_j)) t^{n+j}.$$

But, as we saw above, $e_i(P_n(x)) = 0$ for $i = 0,\ldots,n-1$, that is

$$e_i(\alpha_0 + \alpha_1 x + \ldots + \alpha_{n-1} x^{n-1} + x^n) = 0$$

or

$$\alpha_0 \, L_i(x_0) + \alpha_1 L_i(x_1) + \ldots + \alpha_{n-1} L_i(x_{n-1}) + L_i(x_n) = 0$$

for $i = 0,\ldots,n-1$. Thus the first term in the error cancels if $i \leq n-1$ and we finally have the approximation property

$$L_i(u) \, \tilde{P}_n(t) - \tilde{Q}_n^{(i)}(t) = \left\{ \begin{array}{ll} O(t^{n+1}) & i = 0,\ldots,n-1 \\ O(t^n) & i \geq n. \end{array} \right.$$

The vector Padé approximants of J. Van Iseghem [101] are a particular case of the preceding ones. Their better approximation properties are due to the relations which hold among the functionals e_i (or, equivalently, the functionals L_i)

$$e_{i+md}(x^j) = e_i(x^{j+m}) \qquad\qquad i = 0,\ldots,d-1.$$

The ordinary Padé approximants correspond to $d = 1$.
The results given above generalize the interpretation of Padé approximants due to Hendriksen and Van Rossum [97] which makes use of oblique projection since

$$A_n^n \, x_0 = x_n - P_n(A)x_0 = x_n - (\alpha_0 x_0 + \ldots + \alpha_{n-1} x_{n-1} + x_n) = I_{n-1}(x_n).$$

Let us now consider the Fredholm equation in E^*

$$v = tBv + L_0$$

and the approximate equation

$$v_n = tB_n v_n + L_0$$

where B_n is the operator obtained by the method of moments in E^*. The solution v can be formally written as a Neumann series

$$v = L_0 + tBL_0 + t^2B^2L_0 + ...$$

$$= L_0 + tL_1 + t^2L_2 + ...$$

and we have

$$v(x_i) = L_0(x_i) + L_1(x_i)t + L_2(x_i)t^2 + ... = g_i(t).$$

Let us study $v_n(x_i)$. We have

$$v_n = (I^* - tB_n)^{-1}L_0$$

if t^{-1} is not an eigenvalue of B_n. Thus, we formally have

$$(I^* - tB_n)^{-1} = I^* + tB_n + t^2B_n^2 + ...$$

and

$$v_n(x_i) = L_0(x_i) + B_nL_0(x_i)t + B_n^2 L_0(x_i)t^2 + ...$$

$$= L_0(x_i) + L_1(x_i)t + ... + L_{n-1}(x_i)t^{n-1} + O(t^n)$$

$$= v(x_i) + O(t^n).$$

Let m_i be the linear functional on P defined by

$$m_i(x^j) = L_j(x_i)$$

and let $\{Q_n\}$ be the polynomials obtained by the method of moments in E^*. Then $m_i(Q_n(x)) = 0$ for $i = 0,...,n-1$ which shows that $\{Q_n\}$ is a family of biorthogonal polynomials.
We set

$$V_n^{(i)}(t) = m_i\left(\frac{Q_n(x) - Q_n(t)}{x-t}\right)$$

$$\tilde{Q}_n(t) = t^n Q_n(t^{-1})$$

$$\tilde{V}_n^{(i)}(t) = t^{n-1}V_n^{(i)}(t^{-1}).$$

Then we can prove that

$$v_n(x_i) = \tilde{V}_n^{(i)}(t) \, / \, \tilde{Q}_n(t)$$

and that

$$v(x_i) \, \tilde{Q}_n(t) - \tilde{V}_n^{(i)}(t) = \begin{cases} O(t^{n+1}) \; i = 0,\ldots,n-1 \\ \quad O(t^n) \; i \geq n. \end{cases}$$

Let us set

$$K_n(x,t) = - \begin{vmatrix} 0 & 1 & \ldots & x^k \\ 1 & L_0(x_0) & \ldots & L_0(x_n) \\ \ldots & \ldots & \ldots & \ldots \\ t^n & L_n(x_0) & \ldots & L_n(x_n) \end{vmatrix} \Big/ G_{n+1}.$$

Then it is easy to see that

$$e_i(K_n(x,t)) = m_i(K_n(x,t)) = t^i \qquad \text{for } i = 0,\ldots,n.$$

We previously related $L_i(u_n)$ and $v_n(x_i)$ to Padé-type approximants. Let us now describe a similar relation for u_n and v_n.

We consider again the approximate equation

$$u_n = tA_n u_n + x_0.$$

As shown by Vorobyev [186, p. 28] we have

$$u_n = a_0 x_0 + \ldots + a_{n-1} x_{n-1} = (a_0 + a_1 A + \ldots + a_{n-1} A^{n-1}) x_0$$

with

$$a_0 = 1 - \frac{\alpha_0}{P_n(t^{-1})}$$

$$a_i = t \, a_{i-1} - \frac{\alpha_i}{P_n(t^{-1})} \qquad\qquad i = 1,\ldots,n-1$$

where the α_i's are the coefficients of the polynomial P_n obtained by the method of moments. It can be easily proved by induction that

$$a_i = t^i - \frac{t^n}{\tilde{P}_n(t)} (\alpha_0 t^i + \alpha_1 t^{i-1} + ... + \alpha_i) \qquad\qquad i = 0,...,n-1.$$

But $\tilde{P}_n(t) = \alpha_0 t^n + ... + \alpha_{n-1} t + 1$ and thus we have

$$a_i \tilde{P}_n(t) = t^i (1 + \alpha_{n-1} t + ... + \alpha_{i+1} t^{n-i-1}) \qquad\qquad i = 0,...,n-1.$$

Setting

$$F_n(x) = a_0 + a_1 x + ... + a_{n-1} x^{n-1}$$

it is easy to check that

$$F_n(x)\ \tilde{P}_n(x) = 1 + (\alpha_{n-1} + x)t + (\alpha_{n-2} + \alpha_{n-1}x + x^2)t^2 + ...$$

$$... + (\alpha_1 + \alpha_2 x + ... + \alpha_{n-1}x^{n-2} + x^{n-1})\ t^{n-1}.$$

We set

$$\tilde{Q}_n(x,t) = F_n(x)\ \tilde{P}_n(t)$$

and

$$Q_n(x,t) = t^{n-1}\ \tilde{Q}_n(x,t^{-1}).$$

Then

$$Q_n(x,t) = \frac{P_n(x) - P_n(t)}{x - t} \quad .$$

Thus

$$e_i(\tilde{Q}_{n(x,t)}) = \tilde{Q}_n^{(i)}(t)$$

and

$$e_i(F_n(x)) = \tilde{Q}_n^{(i)}(t) / \tilde{P}_n(t) = L_i(u_n).$$

Then $\tilde{Q}_n(A,t)x_0/\tilde{P}_n(t)$ is a generalization to a vector space of Padé approximants.

We have

$$F_n(A)x_0 = \tilde{Q}_n(A,t)x_0/\tilde{P}_n(t) = a_0x_0 + \dots + a_{n-1}A^{n-1} x_0 = u_n$$

and

$$F_n(A_n) = (I - tA_n)^{-1}.$$

Let $F(x) = (1-xt)^{-1} = 1 + xt + x^2t^2 + \dots$. We formally have

$$F(A)x_0 = x_0 + tAx_0 + t^2A^2x_0 + \dots = u$$

the solution of $u = tAu + x_0$.
We have

$$F(x) \tilde{P}_n(t) = 1 + (\alpha_{n-1} + x)t + (\alpha_{n-2} + \alpha_{n-1}x + x^2)t^2 + \dots$$

$$+ (\alpha_1 + \alpha_2x + \dots + \alpha_{n-1}x^{n-2} + x^{n-1})t^{n-1}$$

$$+ t^n \sum_{j=0}^{\infty} (\alpha_0x^j + \alpha_1x^{j+1} + \dots + \alpha_{n-1}x^{n+j-1} + x^{n+j})t^j.$$

Thus

$$\tilde{P}_n(t)F(A)x_0 = \tilde{Q}_n(A,t)x_0 + O(t^k)$$

that is

$$F_n(A)x_0 = F(A)x_0 + O(t^k).$$

Let us set

$$P(x) = (1-xt)^{-1}(1 - t^nP_n(x)/\tilde{P}_n(t)).$$

P is the Hermite interpolation polynomial of $(1-xt)^{-1}$ at the zeros of P_n. For the usual Padé approximants this is the reason why an approximation in $O(t^{2k})$ is obtained instead of an approximation in $O(t^k)$. This increase in the order of approximation is now lost as we saw before.

We have

$$L_i(P(A)x_0) = L_i((I-tA)^{-1}x_0) - \frac{t^n}{\tilde{P}_n(t)} L_i((I-tA)^{-1}P_n(A)x_0).$$

The interpolation property holds if $\forall_j \geq 0$ and for $i = 0,...,n-1$

$$L_i(A^j P_n(A) x_0) = 0.$$

For $j = 0$ we have

$$L_i(P_n(A) x_0) = L_i(\alpha_0 x_0 + ... + \alpha_{n-1} x_{n-1} + x_n) = L_i(x_n^*) = 0$$

for $i = 0,...,n-1$. But a similar property does not hold for $j > 0$. However, since it is true for $j = 0$, then

$$L_i(P(A) x_0) = L_i ((I-tA)^{-1} x_0) + O(t^{n+1}).$$

Since $P_n(A_n) x_0 = \alpha_0 x_0 + ... + \alpha_{n-1} x_{n-1} + I_{n-1}(x_n) = 0$ we have

$$P(A_n) x_0 = (I-tA_n)^{-1} x_0 = u_n.$$

Of course, similar results can be obtained in E^*.

4 - ADJACENT BIORTHOGONAL FAMILIES

In the previous sections we made use of $x_0,...,x_n$ and $L_0,...,L_n$ to define x_n^*, L_n^*, R_n, M_n and K_n. Of course similar definitions can be given by starting with L_i instead of L_0 (that is using $L_i,...,L_{i+n}$) and with x_j instead of x_0 (that is using $x_j,...,x_{j+n}$). The corresponding elements will be respectively denote by $x_n^{(i,j)}$, $L_n^{(i,j)}$, $R_n^{(i,j)}$, $M_n^{(i,j)}$, and $K_n^{(i,j)}$. The case $i = j = 0$ corresponds to what was done above. The various families $\{L_n^{(i,j)}, x_m^{(i,j)}\}$ obtained for various values of i and j are called adjacent biorthogonal families. The aim of this section is to provide recurrence relations between adjacent $x_n^{(i,j)}$, $L_n^{(i,j)}$, $R_n^{(i,j)}$, $M_n^{(i,j)}$, and $K_n^{(i,j)}$. Such relations will be useful in applications for their practical computation. For this purpose we shall make use of two determinantal identities named after Sylvester and Schweins. They are classical identities which have been recently proved to hold for determinants whose first (or last) row (or column) contains elements of a vector space, all the other entries being scalars [23]. They are given in appendix 3.

The formulae are divided into two classes according whether they involve quantities whose lower indexes can vary by at most one unity (one-step formulae) or more (multistep formulae).

Let us first give some definitions. We set

$$N_{n+1}^{(i,j)} = \begin{vmatrix} L_i(x_j) & \dots & L_i(x_{j+n}) \\ \dots & \dots & \dots \\ L_{i+n-1}(x_j) & \dots & L_{i+n-1}(x_{j+n}) \\ x_j & \dots & x_{j+n} \end{vmatrix}$$

$$D_n^{(i,j)} = \begin{vmatrix} L_i(x_j) & \dots & L_i(x_{j+n-1}) \\ \dots & \dots & \dots \\ L_{i+n-1}(x_j) & \dots & L_{i+n-1}(x_{j+n-1}) \end{vmatrix}$$

$$x_n^{(i,j)} = N_{n+1}^{(i,j)} / D_n^{(i,j)}.$$

We also set

$$\bar{N}_{n+1}^{(i,j)} = \begin{vmatrix} L_i(x_j) & \dots & L_{i+n}(x_j) \\ \dots & \dots & \dots \\ L_i(x_{j+n-1}) & \dots & L_{i+n}(x_{j+n-1}) \\ L_i & \dots & L_{i+n} \end{vmatrix}$$

and

$$L_n^{(i,j)} = \bar{N}_{n+1}^{(i,j)} / D_{n+1}^{(i,j)}.$$

We have

$$L_n^{(i,j)}(x_m^{(i,j)}) = \delta_{nm}$$

$$L_{i+n}(N_{n+1}^{(i,j)}) = \bar{N}_{n+1}^{(i,j)}(x_j) = D_{n+1}^{(i,j)}.$$

Thus

$$L_{i+n}(x_n^{(i,j)}) = D_{n+1}^{(i,j)} / D_n^{(i,j)} \tag{1}$$

$$L_i(x_n^{(i+1,j)}) = (-1)^n D_{n+1}^{(i,j)} / D_n^{(i+1,j)} \tag{2}$$

$$L_n^{(i,j+1)}(x_j) = (-1)^n D_{n+1}^{(i,j)} / D_{n+1}^{(i,j+1)}. \qquad (3)$$

$R_n^{(i,j)}$ is the unique element of Span $(x_0^{(i,j)},...,x_n^{(i,j)})$ satisfying

$$L_p(R_n^{(i,j)}) = w_p \qquad \text{for } p = i,...,i+n$$

where the w_p's are given numbers not all zero which can depend on i and/or j. (Sometimes we shall denote them by $w_p^{(i,j)}$).

$M_n^{(i,j)}$ is the unique element of Span $(L_0^{(i,j)},...,L_n^{(i,j)})$ satisfying

$$M_n^{(i,j)}(x_p) = v_p \qquad \text{for } p = j,...,j+n$$

where the v_p's are given numbers not all zero which can depend on i and/or j (sometimes we shall denote them by $v_p^{(i,j)}$).

Finally let us set

$$N_{n+2}^{(i,j)}(L,f) = \begin{vmatrix} 0 & L_i(f) & ... & L_{i+n}(f) \\ L(x_j) & L_i(x_j) & ... & L_{i+n}(x_j) \\ ... & ... & ... & ... \\ L(x_{j+n}) & L_i(x_{j+n}) & ... & L_{i+n}(x_{j+n}) \end{vmatrix}$$

and

$$K_n^{(i,j)}(L,f) = - N_{n+2}^{(i,j)}(L,f)/D_{n+1}^{(i,j)}.$$

As before we have

$$K_n^{(i,j)}(L,\bullet) = M_n^{(i,j)} \qquad \text{if } v_p = L(x_p)$$

$$K_n^{(i,j)}(\bullet,f) = R_n^{(i,j)} \qquad \text{if } w_p = L_p(f),$$

and

$$K_n^{(i,j)}(L,f) = M_n^{(i,j)}(f) = L(R_n^{(i,j)}).$$

4.1 - One-step formulae.

We shall begin by three basic identities which follow directly from the case $i = j = 0$.

$$\mathbf{F_1} : R_n^{(i,j)} = R_{n-1}^{(i,j)} + \frac{w_{i+n}^{(i,j)} - L_{i+n}(R_{n-1}^{(i,j)})}{L_{i+n}(x_n^{(i,j)})} \, x_n^{(i,j)} \quad \text{with } R_{-1}^{(i,j)} = 0$$

$$\mathbf{F_2} : M_n^{(i,j)} = M_{n-1}^{(i,j)} + [v_{j+n}^{(i,j)} - M_{n-1}^{(i,j)}(x_{j+n})] \, L_n^{(i,j)} \quad \text{with } M_{-1}^{(i,j)} = 0$$

$$\mathbf{F_3} : K_n^{(i,j)}(L,f) = K_{n-1}^{(i,j)}(L,f) + L_n^{(i,j)}(f)L(x_n^{(i,j)}) \quad \text{with } K_{-1}^{(i,j)}(L,f) = 0$$

If we apply Sylvester's identity to $N_{n+1}^{(i,j)}$ we obtain

$$N_{n+1}^{(i,j)} \, D_{n-1}^{(i+1,j+1)} = N_n^{(i+1,j+1)} \, D_n^{(i,j)} - N_n^{(i+1,j)} \, D_n^{(i,j+1)}.$$

Dividing both sides by $D_{n-1}^{(i+1,j+1)} \, D_n^{(i,j)}$, and making use of (2) we get

$$\mathbf{F_4} : x_n^{(i,j)} = x_{n-1}^{(i+1,j+1)} - \frac{L_i(x_{n-1}^{(i+1,j+1)})}{L_i(x_{n-1}^{(i+1,j)})} \, x_{n-1}^{(i+1,j)}$$

Now if we put the last row of $N_{n+1}^{(i,j)}$ as the first one $(N_{n+1}^{(i,j)}$ becomes $(-1)^n N_{n+1}^{(i,j)})$ and if we apply Sylvester's identity, we have

$$N_{n+1}^{(i,j)} D_{n-1}^{(i,j+1)} = N_n^{(i,j+1)} D_n^{(i,j)} - N_n^{(i,j)} D_n^{(i,j+1)}.$$

Dividing both sides by $D_{n-1}^{(i,j+1)} D_n^{(i,j)}$ and using (1), we obtain

$$F_5 : x_n^{(i,j)} = x_{n-1}^{(i,j+1)} - \frac{L_{i+n-1}(x_{n-1}^{(i,j+1)})}{L_{i+n-1}(x_{n-1}^{(i,j)})} x_{n-1}^{(i,j)}$$

Let us now apply Schweins' identity to $(-1)^n N_{n+1}^{(i,j)}$. We obtain

$$(-1)^n N_{n+1}^{(i,j)} D_n^{(i+1,j)} = (-1)^n N_{n+1}^{(i+1,j)} D_n^{(i,j)} - (-1)^{n-1} N_n^{(i+1,j)} D_{n+1}^{(i,j)}.$$

Dividing by $D_n^{(i+1,j)} D_n^{(i,j)}$ and using (2) we get

$$F_6 : x_n^{(i,j)} = x_n^{(i+1,j)} - \frac{L_i(x_n^{(i+1,j)})}{L_i(x_{n-1}^{(i+1,j)})} x_{n-1}^{(i+1,j)}$$

Using (1) instead of (2) leads to

$$F_7 : x_n^{(i+1,j)} = x_n^{(i,j)} - \frac{L_{i+n}(x_n^{(i,j)})}{L_{i+n}(x_{n-1}^{(i+1,j)})} x_{n-1}^{(i+1,j)}$$

Let us now give similar relations for the $L_n^{(i,j)}$'s.

If we apply Sylvester's identity to $\bar{N}_{n+1}^{(i,j)}$, we obtain

$$\bar{N}_{n+1}^{(i,j)} \, D_{n-1}^{(i+1,j+1)} = \bar{N}_n^{(i+1,j+1)} \, D_n^{(i,j)} - \bar{N}_n^{(i,j+1)} \, D_n^{(i+1,j)}.$$

Dividing by $D_{n-1}^{(i+1,j+1)} \, D_{n+1}^{(i,j)}$, we obtain

$$L_n^{(i,j)} = L_{n-1}^{(i+1,j+1)} \, \frac{D_n^{(i,j)} D_n^{(i+1,j+1)}}{D_{n-1}^{(i+1,j+1)} \, D_{n+1}^{(i,j)}} - L_{n-1}^{(i,j+1)} \frac{D_n^{(i+1,j)} D_n^{(i,j+1)}}{D_{n-1}^{(i+1,j+1)} D_{n+1}^{(i,j)}} \, .$$

Now if we apply Sylvester's identity to $D_{n+1}^{(i,j)}$

$$D_{n+1}^{(i,j)} \, D_{n-1}^{(i+1,j+1)} = D_n^{(i,j)} \, D_n^{(i+1,j+1)} - D_n^{(i,j+1)} \, D_n^{(i+1,j)}$$

and then make use of (3) we obtain

$$\frac{D_{n+1}^{(i,j)} \, D_{n-1}^{(i+1,j+1)}}{D_n^{(i,j)} \, D_n^{(i+1,j+1)}} = 1 - \frac{D_n^{(i,j+1)} \, D_n^{(i+1,j)}}{D_n^{(i,j)} \, D_n^{(i+1,j+1)}} = 1 - \frac{L_{n-1}^{(i+1,j+1)}(x_j)}{L_{n-1}^{(i,j+1)}(x_j)}$$

$$\frac{D_{n+1}^{(i,j)} \, D_{n-1}^{(i+1,j+1)}}{D_n^{(i,j+1)} \, D_n^{(i+1,j)}} = \frac{D_n^{(i,j)} \, D_n^{(i+1,j+1)}}{D_n^{(i,j+1)} \, D_n^{(i+1,j)}} - 1 = \frac{L_{n-1}^{(i,j+1)}(x_j)}{L_{n-1}^{(i+1,j+1)}(x_j)} - 1$$

and thus we finally have

$$\boxed{F_8 : L_n^{(i,j)} = \frac{L_{n-1}^{(i,j+1)}(x_j) L_{n-1}^{(i+1,j+1)} - L_{n-1}^{(i+1,j+1)}(x_j) L_{n-1}^{(i,j+1)}}{L_{n-1}^{(i,j+1)}(x_j) - L_{n-1}^{(i+1,j+1)}(x_j)}}$$

In $\bar{N}_{n+1}^{(i,j)}$ let us put the last row as the first one (we obtain $(-1)^n \, \bar{N}_{n+1}^{(i,j)}$) and then apply Sylvester's identity

$$\bar{N}_{n+1}^{(i,j)} \, D_{n-1}^{(i+1,j)} = \bar{N}_n^{(i+1,j)} \, D_n^{(i,j)} - \bar{N}_n^{(i,j)} \, D_n^{(i+1,j)}.$$

Dividing by $D_{n-1}^{(i+1,j)} \, D_{n+1}^{(i,j)}$ we get

$$L_n^{(i,j)} = (L_{n-1}^{(i+1,j)} - L_{n-1}^{(i,j)}) \frac{D_n^{(i,j)} \, D_n^{(i+1,j)}}{D_{n-1}^{(i+1,j)} \, D_{n+1}^{(i,j)}}.$$

In $D_{n+1}^{(i,j)}$ let us make a transposition, put the last row as the first one (it becomes $(-1)^n \, D_{n+1}^{(i,j)}$) and then apply Sylvester's identity

$$(-1)^n D_{n+1}^{(i,j)} \, D_{n-1}^{(i+1,j)} = (-1)^{n-1}\bar{N}_n^{(i,j)}(x_{j+n}) \, D_n^{(i+1,j)} - (-1)^{n-1} \, \bar{N}_n^{(i+1,j)}(x_{j+n}) \, D_n^{(i,j)} .$$

Thus

$$\frac{D_{n+1}^{(i,j)} \, D_{n-1}^{(i+1,j)}}{D_n^{(i,j)} \, D_n^{(i+1,j)}} = \frac{\bar{N}_n^{(i+1,j)}(x_{j+n})}{D_n^{(i+1,j)}} - \frac{\bar{N}_n^{(i,j)}(x_{j+n})}{D_n^{(i,j)}}$$

$$= L_{n-1}^{(i+1,j)}(x_{j+n}) - L_{n-1}^{(i,j)}(x_{j+n})$$

and we finally obtain

$$\boxed{F_9 : L_n^{(i,j)} = \frac{L_{n-1}^{(i+1,j)} - L_{n-1}^{(i,j)}}{L_{n-1}^{(i+1,j)}(x_{j+n}) - L_{n-1}^{(i,j)}(x_{j+n})}}$$

Let us now put the last row of $\bar{N}_{n+1}^{(i,j)}$ as the first one (thus obtaining $(-1)^n\bar{N}_{n+1}^{(i,j)}$) and apply Schweins' identity

$$(-1)^n \, \bar{N}_{n+1}^{(i,j)} \, D_n^{(i,j+1)} = (-1)^n\bar{N}_{n+1}^{(i,j+1)} \, D_n^{(i,j)} - (-1)^{n-1}\bar{N}_n^{(i,j+1)} \, D_{n+1}^{(i,j)} .$$

Dividing by $D_n^{(i,j+1)} D_{n+1}^{(i,j)}$ we get

$$L_n^{(i,j)} = \frac{D_n^{(i,j)} D_{n+1}^{(i,j+1)}}{D_n^{(i,j+1)} D_{n+1}^{(i,j)}} L_n^{(i,j+1)} + L_{n-1}^{(i,j+1)}.$$

Using (3) we obtain

$$\boxed{F_{10} : L_n^{(i,j)} = L_{n-1}^{(i,j+1)} - \frac{L_{n-1}^{(i,j+1)}(x_j)}{L_n^{(i,j)}(x_j)} L_n^{(i,j+1)}}$$

Let us now give recursive formulae for the $K_n^{(i,j)}$'s.

In F_1 let us put $w_{i+n}^{(i,j)} = w_{i+n} = L_{i+n}(f) = L_{i+n}(R_n^{(i,j)})$.

Moreover $K_n^{(i,j)}(L,f) = L(R_n^{(i,j)})$ and F_1 becomes

$$\boxed{F_{11} : K_n^{(i,j)}(L,f) = K_{n-1}^{(i,j)}(L,f) + \frac{L_{i+n}(R_n^{(i,j)}) - L_{i+n}(R_{n-1}^{(i,j)})}{L_{i+n}(x_n^{(i,j)})} L(x_n^{(i,j)})}$$

In F_2 let us put $v_{j+n}^{(i,j)} = v_{j+n} = L(x_{j+n}) = M_n^{(i,j)}(x_{j+n})$.

Moreover $K_n^{(i,j)}(L,f) = M_n^{(i,j)}(f)$ and F_2 becomes

$$\boxed{F_{12} : K_n^{(i,j)}(L,f) = K_{n-1}^{(i,j)}(L,f) + [M_n^{(i,j)}(x_{j+n}) - M_{n-1}^{(i,j)}(x_{j+n})] L_n^{(i,j)}(f)}$$

We shall now apply Schweins' formula to $N_{n+2}^{(i,j)}(L,f)$. We get

$$N_{n+2}^{(i,j)}(L,f)(-1)^n \, \overline{N}_{n+1}^{(i+1,j)}(f) = N_{n+2}^{(i+1,j)}(L,f)(-1)^n \, \overline{N}_{n+1}^{(i,j)}(f)$$
$$- N_{n+1}^{(i+1,j)}(L,f)(-1)^{n+1} \, \overline{N}_{n+2}^{(i,j)}(f).$$

Dividing by $D_{n+1}^{(i,j)} \, D_{n+1}^{(i+1,j)}$ we obtain

$$\frac{N_{n+2}^{(i,j)}(L,f)}{D_{n+1}^{(i,j)}} \, \frac{\overline{N}_{n+1}^{(i+1,j)}(f)}{D_{n+1}^{(i+1,j)}} = \frac{N_{n+2}^{(i+1,j)}(L,f)}{D_{n+1}^{(i+1,j)}} \, \frac{\overline{N}_{n+1}^{(i,j)}(f)}{D_{n+1}^{(i,j)}}$$

$$+ \frac{N_{n+1}^{(i+1,j)}(L,f)}{D_n^{(i+1,j)}} \, \frac{\overline{N}_{n+2}^{(i,j)}(f)}{D_{n+2}^{(i,j)}} \, \frac{D_n^{(i+1,j)} D_{n+2}^{(i,j)}}{D_{n+1}^{(i,j)} D_{n+1}^{(i+1,j)}} \ .$$

But as we saw before

$$\frac{D_n^{(i+1,j)} D_{n+2}^{(i,j)}}{D_{n+1}^{(i,j)} D_{n+1}^{(i+1,j)}} = L_n^{(i+1,j)}(x_{j+n+1}) - L_n^{(i,j)}(x_{j+n+1})$$

and thus we have

$$K_n^{(i,j)}(L,f) \, L_n^{(i+1,j)}(f) = K_n^{(i+1,j)}(L,f) \, L_n^{(i,j)}(f)$$
$$- [L_n^{(i,j)}(x_{j+n+1}) - L_n^{(i+1,j)}(x_{j+n+1})] \, K_{n-1}^{(i+1,j)}(L,f) \, L_{n+1}^{(i,j)}(f).$$

and then, by F_9

$$[L_n^{(i,j)}(x_{j+n+1}) - L_n^{(i+1,j)}(x_{j+n+1})] L_{n+1}^{(i,j)}(f) = L_n^{(i,j)}(f) - L_n^{(i+1,j)}(f).$$

Thus

$$K_n^{(i,j)}(L,f) \, L_n^{(i+1,j)}(f) = K_n^{(i+1,j)}(L,f) \, L_n^{(i,j)}(f)$$
$$- K_{n-1}^{(i+1,j)}(L,f)[L_n^{(i,j)}(f) - L_n^{(i+1,j)}(f)].$$

But

$$K_n^{(i+1,j)}(L,f) = K_{n-1}^{(i+1,j)}(L,f) + L_n^{(i+1,j)}(f) \, L(x_n^{(i+1,j)}) \text{ and we finally obtain}$$

$$F_{13} : K_n^{(i,j)}(L,f) = K_{n-1}^{(i+1,j)}(L,f) + L_n^{(i,j)}(f) \, L(x_n^{(i+1,j)})$$

Let us now transpose $N_{n+2}^{(i,j)}(L,f)$ and apply Schweins' formula

$$N_{n+2}^{(i,j)}(L,f) \, L(N_{n+1}^{(i,j+1)}) = N_{n+2}^{(i,j+1)}(L,f) \, L(N_{n+1}^{(i,j)}) + N_{n+1}^{(i,j+1)}(L,f) \, L(N_{n+2}^{(i,j)}).$$

Dividing by $D_{n+1}^{(i,j)} \, D_n^{(i,j+1)}$ we have

$$K_n^{(i,j)}(L,f) \, L(x_n^{(i,j+1)}) = K_n^{(i,j+1)}(L,f) \, L(x_n^{(i,j)}) \frac{D_{n+1}^{(i,j+1)} \, D_n^{(i,j)}}{D_{n+1}^{(i,j)} \, D_n^{(i,j+1)}}$$
$$+ K_{n-1}^{(i,j+1)}(L,f) \, L(x_{n+1}^{(i,j)}).$$

But, by (1)

$$\frac{D_{n+1}^{(i,j+1)} \, D_n^{(i,j)}}{D_n^{(i,j+1)} \, D_{n+1}^{(i,j)}} = \frac{L_{i+n}(x_n^{(i,j+1)})}{L_{i+n}(x_n^{(i,j)})}$$

and, by F_5

$$\frac{L_{i+n}(x_n^{(i,j+1)})}{L_{i+n}(x_n^{(i,j)})} \, L(x_n^{(i,j)}) = L(x_n^{(i,j+1)}) - L(x_{n+1}^{(i,j)}).$$

Thus we finally obtain

$$F_{14} : K_n^{(i,j)}(L,f) = K_n^{(i,j+1)}(L,f) - L_n^{(i,j+1)}(f)\, L(x_{n+1}^{(i,j)})$$

Using F_5 this relation becomes

$$F_{15} : K_n^{(i,j)}(L,f) = K_{n-1}^{(i,j+1)}(L,f) + L_n^{(i,j+1)}(f)L(x_n^{(i,j)})\, \frac{L_{i+n}(x_n^{(i,j+1)})}{L_{i+n}(x_n^{(i,j)})}$$

Using F_3, F_{13} gives

$$K_n^{(i,j)}(L,f) = K_n^{(i+1,j)}(L,f) + L(x_n^{(i+1,j)})[L_n^{(i,j)}(f) - L_n^{(i+1,j)}(f)]$$

and, from F_9, we obtain

$$F_{16} : K_n^{(i,j)}(L,f) = K_n^{(i+1,j)}(L,f) + L(x_n^{(i+1,j)})\, L_{n+1}^{(i,j)}(f)[L_n^{(i,j)}(x_{j+n+1})$$
$$-L_n^{(i+1,j)}(x_{j+n+1})]$$

From F_3 we obtain the expression of $L_n^{(i,j)}(f)$ and replace it in F_{13}. Thus we have

$$F_{17} : K_n^{(i,j)}(L,f) = \frac{L(x_n^{(i,j)})K_{n-1}^{(i+1,j)}(L,f) - L(x_n^{(i+1,j)})K_{n-1}^{(i,j)}(L,f)}{L(x_n^{(i,j)}) - L(x_n^{(i+1,j)})}$$

Similarly F_{15} becomes by replacing $L(x_n^{(i,j)})$ by its expression from F_3

$$F_{18} : K_n^{(i,j)}(L,f) = \frac{L_n^{(i,j)}(f) L_{i+n}(x_n^{(i,j)}) K_{n-1}^{(i,j+1)}(L,f) - L_n^{(i,j+1)}(f) L_{i+n}(x_n^{(i,j+1)}) K_{n-1}^{(i,j)}(L,f)}{L_n^{(i,j)}(f) L_{i+n}(x_n^{(i,j)}) - L_n^{(i,j+1)}(f) L_{i+n}(x_n^{(i,j+1)})}$$

Equating F_3 and F_{17} we find

$$K_{n-1}^{(i+1,j)}(L,f) = K_{n-1}^{(i,j)}(L,f) + L_n^{(i,j)}(f) \, [L(x_n^{(i,j)}) - L(x_n^{(i+1,j)})].$$

Using F_6, this relation becomes

$$F_{19} : K_n^{(i,j)}(L,f) = K_n^{(i+1,j)}(L,f) + L(x_n^{(i+1,j)})L_{n+1}^{(i,j)}(f)\frac{L_i(x_{n+1}^{(i+1,j)})}{L_i(x_n^{(i+1,j)})}$$

Since $K_n^{(i,j)}(L,\bullet) = M_n^{(i,j)}$ and $K_n^{(i,j)}(\bullet,f) = R_n^{(i,j)}$ the preceding relations for the $K_n^{(i,j)}$'s give relations for the $M_n^{(i,j)}$'s and the $R_n^{(i,j)}$'s.

From F_{13} have

$$F_{20} : R_n^{(i,j)} = R_{n-1}^{(i+1,j)} + L_n^{(i,j)}(f)\, x_n^{(i+1,j)}$$

$$F_{21} : M_n^{(i,j)} = M_{n-1}^{(i+1,j)} + L(x_n^{(i+1,j)})L_n^{(i,j)}$$

From F_{14} we have

$$F_{22} : R_n^{(i,j)} = R_n^{(i,j+1)} - L_n^{(i,j+1)}(f)\, x_{n+1}^{(i,j)}$$

$$F_{23} : M_n^{(i,j)} = M_n^{(i,j+1)} - L(x_{n+1}^{(i,j)})\, L_n^{(i,j+1)}$$

From F_{15} it follows

$$F_{24} : R_n^{(i,j)} = R_{n-1}^{(i,j+1)} + \frac{L_{i+n}(x_n^{(i,j+1)})}{L_{i+n}(x_n^{(i,j)})}\, L_n^{(i,j+1)}(f)x_n^{(i,j)}$$

$$F_{25} : M_n^{(i,j)} = M_{n-1}^{(i,j+1)} + \frac{L_{i+n}(x_n^{(i,j+1)})}{L_{i+n}(x_n^{(i,j)})}\, L(x_n^{(i,j)})L_n^{(i,j+1)}$$

From F_{16} we obtain

$$F_{26} : R_n^{(i,j)} = R_n^{(i+1,j)} + [L_n^{(i,j)}(x_{j+n+1}) - L_n^{(i+1,j)}(x_{j+n+1})] \, L_{n+1}^{(i,j)}(f) x_n^{(i+1,j)}$$

$$F_{27} : M_n^{(i,j)} = M_n^{(i+1,j)} + [L_n^{(i,j)}(x_{j+n+1}) - L_n^{(i+1,j)}(x_{j+n+1})] \, L(x_n^{(i+1,j)}) L_{n+1}^{(i,j)}$$

From F_{17} we find

$$F_{28} : M_n^{(i,j)} = \frac{L(x_n^{(i,j)})M_{n-1}^{(i+1,j)} - L(x_n^{(i+1,j)})M_{n-1}^{(i,j)}}{L(x_n^{(i,j)}) - L(x_n^{(i+1,j)})}$$

From F_{18} we have

$$F_{29} : R_n^{(i,j)} =$$

$$\frac{L_n^{(i,j)}(f)L_{i+n}(x_n^{(i,j)})R_{n-1}^{(i,j+1)} - L_n^{(i,j+1)}(f)L_{i+n}(x_n^{(i,j+1)})R_{n-1}^{(i,j)}}{L_n^{(i,j)}(f)L_{i+n}(x_n^{(i,j)}) - L_n^{(i,j+1)}(f)L_{i+n}(x_n^{(i,j+1)})}$$

From F_{19} it follows

$$F_{30} : R_n^{(i,j)} = R_n^{(i+1,j)} + \frac{L_i(x_{n+1}^{(i+1,j)})}{L_i(x_n^{(i+1,j)})} \, L_{n+1}^{(i,j)}(f) x_n^{(i+1,j)}$$

$$F_{31} : M_n^{(i,j)} = M_n^{(i+1,j)} + \frac{L_i(x_{n+1}^{(i+1,j)})}{L_i(x_n^{(i+1,j)})} L(x_{n+1}^{(i+1,j)})L_{n+1}^{(i,j)}$$

Let us now generalize the well-known divided differences. Using F_1 to generalize the results given in section 3.2 we immediately see that

$$R_n^{(i,j)} = R_{n-1}^{(i,j)} + a_n^{(i,j)} \, x_n^{(i,j)}$$

with

$$a_n^{(i,j)} = \begin{vmatrix} L_i(x_j) & ... & L_{i+n}(x_j) \\ ... & ... & ... \\ L_i(x_{j+n-1}) & ... & L_{i+n}(x_{j+n-1}) \\ L_i(f) & ... & L_{i+n}(f) \end{vmatrix} / D_{n+1}^{(i,j)}.$$

$a_n^{(i,j)}$ is a generalization of divided differences and we shall now make use of the notation

$$a_n^{(i,j)} = \begin{bmatrix} x_j...x_{j+n} \\ L_i...L_{i+n} \end{bmatrix} \Big| f \Big] = \begin{vmatrix} L_i(x_j) & ... & L_{i+n}(x_j) \\ ... & ... & ... \\ L_i(x_{j+n-1}) & ... & L_{i+n}(x_{j+n-1}) \\ L_i(f) & ... & L_{i+n}(f) \end{vmatrix} / D_{n+1}^{(i,j)}.$$

From F_1 we see that

$$\begin{bmatrix} x_j...x_{j+n} \\ L_i...L_{i+n} \end{bmatrix} \Big| f \Big] = \frac{L_{i+n}(f - R_{n-1}^{(i,j)})}{L_{i+n}(x_n^{(i,j)})}.$$

Thus

$$\begin{bmatrix} x_j...x_{j+n} \\ L_i...L_{i+n} \end{bmatrix} \Big| x_p \Big] = \begin{cases} 0 & p = j,...,j+n-1 \\ 1 & p = j+n. \end{cases}$$

We also have

$$L_n^{(i,j)}(f) = \begin{bmatrix} x_j...x_{j+n} \\ L_i...L_{i+n} \end{bmatrix} f$$

and thus $a_n^{(i,j)} = L_n^{(i,j)}(f)$.

If follows that the recurrence relations between the $L_n^{(i,j)}$'s (that is F_8, F_9 and F_{10}) give recurrence relations for computing the preceding generalized divided differences. In particular F_9 is the well-known recurrence relation which was generalized by Mühlbach [139] for interpolation by a linear combination of functions forming a Chebyshev system (that is such that the denominators do not vanish). We have now generalized further this formula.

Of course similar results hold in E^*. We have

$$M_n^{(i,j)} = M_{n-1}^{(i,j)} + b_n^{(i,j)} L_n^{(i,j)}$$

with

$$b_n^{(i,j)} = \begin{vmatrix} L_i(x_j) & ... & L_i(x_{j+n}) \\ ... & ... & ... \\ L_{i+n-1}(x_j) & ... & L_{i+n-1}(x_{j+n}) \\ L(x_j) & ... & L(x_{j+n}) \end{vmatrix} / D_n^{(i,j)}.$$

We shall set

$$b_n^{(i,j)} = \begin{bmatrix} L_i & ... & L_{i+n-1} \\ x_j & ... & x_{j+n-1} \;\; x_{j+n} \end{bmatrix} L$$

and call it a dual divided difference.

By F_2 we have

$$(L - M_{n-1}^{(i,j)})(x_{j+n}) = \begin{bmatrix} L_i & ... & L_{i+n-1} \\ x_j & ... & x_{j+n-1} \;\; x_{j+n} \end{bmatrix} L.$$

Moreover

$$\begin{bmatrix} L_i & \dots & L_{i+n-1} \\ x_j & \dots & x_{j+n-1} \ x_{j+n} \end{bmatrix} L_p \end{bmatrix} = \begin{cases} 0 & p = i,\dots,i+n-1 \\ D_{n+1}^{(i,j)}/D_n^{(i,j)} & p = i+n. \end{cases}$$

We also have

$$L(x_n^{(i,j)}) = \begin{bmatrix} L_i & \dots & L_{i+n-1} \\ x_j & \dots & x_{j+n-1} \ x_{j+n} \end{bmatrix} L \end{bmatrix}$$

and thus $b_n^{(i,j)} = L(x_n^{(i,j)})$.

It follows that the recurrence relations for the $x_n^{(i,j)}$'s (that is F_4, F_5, F_6 and F_7) provide recurrence relations for the dual divided differences.

We see that we also have

$$R_n^{(i,j)} = \sum_{k=o}^{n} L_k^{(i,j)}(f) x_k^{(i,j)}$$

$$M_n^{(i,j)} = \sum_{k=o}^{n} L(x_k^{(i,j)}) L_k^{(i,j)}$$

$$L(R_n^{(i,j)}) = M_n^{(i,j)}(f) = K_n^{(i,j)}(L,f) = \sum_{k=o}^{n} L(x_k^{(i,j)}) \ L_k^{(i,j)}(f).$$

Let e be an arbitrary linear functional. Then

$$e(f - R_{n-1}^{(i,j)}) = e(x_n^{(i,j)}) \begin{bmatrix} x_j & \dots & x_{j+n-1} \ x_{j+n} \\ L_i & \dots & L_{i+n-1} \ e \end{bmatrix} f \end{bmatrix}.$$

Similarly let p be an arbitrary element of E. Then

$$(L - M_{n-1}^{(i,j)})(p) = \begin{bmatrix} L_i & \dots & L_{i+n-1} \\ x_j & \dots & x_{j+n-1} \ p \end{bmatrix} L \end{bmatrix}.$$

Thus we have obtained generalizations of well known expressions for the interpolation error in terms of divided differences and dual divided differences.

These generalized divided differences can be used for Hermite interpolation in $\mathbb{R}^k$ thus leading to a generalization of Newton's recursive interpolation as presented in [94] (compare with [147]). They also have applications in recurrence relations for Chebyshevian B-splines [130]. It is also clear from formula (38b) of [134] that the preceding recursive formulae could be useful for the algorithmic aspects of surface spline interpolation, a point mentioned as been not "widely known in spite of its fundamental simplicity". Other applications to splines were given in [122]. All these points and connections deserve further research and, in particular, the recursive algorithms must be written in full and their numerical stability must be studied.

4.2 - Multistep formulae

Let us now consider the following ratios of determinants

$$
H_k^{(n)} = \begin{vmatrix} e_n & \dots & e_{n+k} \\ g_1(n) & \dots & g_1(n+k) \\ \dots & \dots & \dots \\ g_k(n) & \dots & g_k(n+k) \end{vmatrix} \bigg/ \begin{vmatrix} 1 & \dots & 1 \\ g_1(n) & \dots & g_1(n+k) \\ \dots & \dots & \dots \\ g_k(n) & \dots & g_k(n+k) \end{vmatrix}
$$

$$
g_{k,i}^{(n)} = \begin{vmatrix} g_i(n) & \dots & g_i(n+k) \\ g_1(n) & \dots & g_1(n+k) \\ \dots & \dots & \dots \\ g_k(n) & \dots & g_k(n+k) \end{vmatrix} \bigg/ \begin{vmatrix} 1 & \dots & 1 \\ g_1(n) & \dots & g_1(n+k) \\ \dots & \dots & \dots \\ g_k(n) & \dots & g_k(n+k) \end{vmatrix}
$$

where the e's are elements of a vector space and the $g_i(j)$ are scalars.

Applying Sylvester's identity to the numerators of $H_k^{(n)}$ and $g_{k,i}^{(n)}$ and to their common denominator immediately leads to the following recursive scheme known as the H-algorithm [22, 40] :

$$
H_0^{(n)} = e_n \qquad\qquad g_{0,i}^{(n)} = g_i(n)
$$

$$H_k^{(n)} = \frac{g_{k-1,k}^{(n+1)} H_{k-1}^{(n)} - g_{k-1,k}^{(n)} H_{k-1}^{(n+1)}}{g_{k-1,k}^{(n+1)} - g_{k-1,k}^{(n)}} \qquad k = 1,2,\ldots;\ n = 0,1,\ldots$$

$$g_{k,i}^{(n)} = \frac{g_{k-1,k}^{(n+1)} g_{k-1,i}^{(n)} - g_{k-1,k}^{(n+1)} g_{k-1,i}^{(n+1)}}{g_{k-1,k}^{(n+1)} - g_{k-1,k}^{(n)}} \qquad k = 1,2,\ldots;\ n = 0,1,\ldots\ ;\ i = k+1,\ldots$$

Let us now apply the H-algorithm with the initializations

$$\tilde{H}_0^{(n)} = H_m^{(n)} \qquad \tilde{g}_{0,i}^{(n)} = g_{m,m+i}^{(n)}$$

and let us denote by $\tilde{H}_k^{(n)}$ and $\tilde{g}_{k,i}^{(n)}$ the results thus obtained. But, since the H-algorithm involves a recursion on the index k, we have

$$\tilde{H}_k^{(n)} = H_{k+m}^{(n)} \qquad \text{and} \qquad \tilde{g}_{k,i}^{(n)} = g_{k+m,k+m+i}^{(n)} .$$

On the other hand $\tilde{H}_k^{(n)}$ and $\tilde{g}_{k,i}^{(n)}$ are given by ratios of determinants similar to those for $H_k^{(n)}$ and $g_{k,i}^{(n)}$ but with the new initializations $H_m^{(n)}$ and $g_{m,m+i}^{(n)}$ instead of $H_0^{(n)}$ and $g_{0,i}^{(n)}$. Thus we have proved that

$$H_{k+m}^{(n)} = \begin{vmatrix} H_m^{(n)} & \ldots & H_m^{(n+k)} \\ g_{m,m+1}^{(n)} & \cdots & g_{m,m+1}^{(n+k)} \\ \ldots & \ldots & \ldots \\ g_{m,m+k}^{(n)} & \cdots & g_{m,m+k}^{(n+k)} \end{vmatrix} \Bigg/ \begin{vmatrix} 1 & \ldots & 1 \\ g_{m,m+1}^{(n)} & \cdots & g_{m,m+1}^{(n+k)} \\ \ldots & \ldots & \ldots \\ g_{m,m+k}^{(n)} & \cdots & g_{m,m+k}^{(n+k)} \end{vmatrix} .$$

Interchanging m and k, we also have

$$H_{k+m}^{(n)} = \begin{vmatrix} H_k^{(n)} & \ldots & H_k^{(n+m)} \\ g_{k,k+1}^{(n)} & \cdots & g_{k,k+1}^{(n+m)} \\ \ldots & \ldots & \ldots \\ g_{k,k+m}^{(n)} & \cdots & g_{k,k+m}^{(n+m)} \end{vmatrix} \Bigg/ \begin{vmatrix} 1 & \ldots & 1 \\ g_{k,k+1}^{(n)} & \cdots & g_{k,k+1}^{(n+m)} \\ \ldots & \ldots & \ldots \\ g_{k,k+m}^{(n)} & \cdots & g_{k,k+m}^{(n+m)} \end{vmatrix} .$$

Similar relations hold for $g_{k+m,k+m+i}^{(n)}$ by replacing the first row of the numerators respectively by $(g_{m,m+i}^{(n)},...,g_{m,m+i}^{(n+k)})$ and $(g_{k,k+i}^{(n)},...,g_{k,k+i}^{(n+m)})$.

If we set $m = 0$ in the first relation we recover the definition of $H_k^{(n)}$ (and $g_{k,i}^{(n)}$) as given above. In the second relation the choice $m = 1$ leads to the H-algorithm. An arbitrary choice of m (in the first formula) or k (in the second formula) gives a recursive method for computing the $H_{k+m}^{(n)}$'s directly in terms of the $H_m^{(n)}$'s (or the $H_k^{(n)}$'s) without computing the intermediate quantities. Such a procedure can be useful when a singularity occurs, that is a division by zero. Such multistep formulae were already proposed in [19, 31] in a less general setting.

We also have

$$
H_{k+m}^{(n)} =
\begin{vmatrix}
H_k^{(n)} & \Delta H_k^{(n)} & ... & \Delta H_k^{(n+m-1)} \\
g_{k,k+1}^{(n)} & \Delta g_{k,k+1}^{(n)} & ... & \Delta g_{k,k+1}^{(n+m-1)} \\
... & ... & ... & ... \\
g_{k,k+m}^{(n)} & \Delta g_{k,k+m}^{(n)} & ... & \Delta g_{k,k+m}^{(n+m-1)}
\end{vmatrix}
\Bigg/
\begin{vmatrix}
\Delta g_{k,k+1}^{(n)} & ... & \Delta g_{k,k+1}^{(n+m-1)} \\
... & ... & ... \\
\Delta g_{k,k+m}^{(n)} & ... & \Delta g_{k,k+m}^{(n+m-1)}
\end{vmatrix}
$$

and a similar relation for $g_{k+m,k+m+i}^{(n)}$ (where Δ acts on the superscripts).

Applying the extension of Schur complement and formula to a vector space as given in [28] we have

$$
H_{k+m}^{(n)} =
H_k^{(n)} - (\Delta H_k^{(n)},...,\Delta H_k^{(n+m-1)}) *
\begin{pmatrix}
\Delta g_{k,k+1}^{(n)} & ... & \Delta g_{k,k+1}^{(n+m-1)} \\
... & ... & ... \\
\Delta g_{k,k+m}^{(n)} & ... & \Delta g_{k,k+m}^{(n+m-1)}
\end{pmatrix}^{-1}
\begin{pmatrix}
g_{k,k+1}^{(n)} \\
... \\
g_{k,k+m}^{(n)}
\end{pmatrix}.
$$

The advantage of this formula over the preceding determinantal formula is that it replaces the computation of determinants by the solution of a system of linear equations. For $k = 0$ we obtain a Nuttall-type formula for the H-algorithm [149].

From the computational point of view, since the Schur complement is related to the bordering method for solving recursively a system of linear equations whose dimension grows, the preceding techniques can linear equations whose dimension grows, the preceding techniques can be used to compute the sequence $(H_k^{(o)})$. Once this sequence has been obtained (and also similarly the sequence $(g_{k,k+1}^{(o)})$) the other $H_k^{(n)}$'s can be obtained by the so-called progressive form of the algorithm [32] whose stability properties are usually better

$$H_{k-1}^{(n+1)} = H_k^{(n)} - \frac{g_{k-1,k}^{(n+1)}}{g_{k-1,k}^{(n)}} [H_k^{(n)} - H_{k-1}^{(n)}].$$

The $g_{k-1,k}^{(n+1)}$ have to be computed by using a special trick, see [39].

We shall now see how to fit some of our previous determinantal formulae into this framework. Thus the one-step or the multistep H-algorithm will provide new procedures for their computation.

We saw that

$$M_k^{(i,j)} = \begin{vmatrix} L(x_j) & L_i(x_j) & ... & L_{i+k}(x_j) \\ 0 & L_i & ... & L_{i+k} \\ L(x_{j+1}) & L_i(x_{j+1}) & ... & L_{i+k}(x_{j+1}) \\ ... & ... & ... & ... \\ L(x_{j+k}) & L_i(x_{j+k}) & ... & L_{i+k}(x_{j+k}) \end{vmatrix} \Bigg/ \begin{vmatrix} L_i(x_j) & ... & L_{i+k}(x_j) \\ ... & ... & ... \\ L_i(x_{j+k}) & ... & L_{i+k}(x_{j+k}) \end{vmatrix}.$$

We shall now prove that we also have

$$M_k^{(i,j)} = \begin{vmatrix} M_0^{(i,j)} & ... & M_0^{(i+k,j)} \\ f_{j+1}^{(i,j)} & ... & f_{j+1}^{(i+k,j)} \\ ... & ... & ... \\ f_{j+k}^{(i,j)} & ... & f_{j+k}^{(i+k,j)} \end{vmatrix} \Bigg/ \begin{vmatrix} 1 & ... & 1 \\ f_{j+1}^{(i,j)} & ... & f_{j+1}^{(i+k,j)} \\ ... & ... & ... \\ f_{j+k}^{(i,j)} & ... & f_{j+k}^{(i+k,j)} \end{vmatrix}$$

with $M_0^{(i,j)} = \dfrac{L(x_j)}{L_i(x_j)} L_i$ and $f_p^{(i,j)} = L_i(x_p) \dfrac{L(x_j)}{L_i(x_j)} - L(x_p)$.

The above numerator can be written as

$$
\begin{vmatrix}
1 & 1 & \cdots & 1 \\
0 & \dfrac{L(x_j)}{L_i(x_j)}L_i & \cdots & \dfrac{L(x_j)}{L_{i+k}(x_j)}L_{i+k} \\
0 & \dfrac{L(x_j)}{L_i(x_j)}L_i(x_{j+1})\text{-}L(x_{j+1}) & \cdots & \dfrac{L(x_j)}{L_{i+k}(x_j)}L_{i+k}(x_{j+1})\text{-}L(x_{j+1}) \\
\cdots & \cdots & \cdots & \cdots \\
0 & \dfrac{L(x_j)}{L_i(x_j)}L_i(x_{j+k})\text{-}L(x_{j+k}) & \cdots & \dfrac{L(x_j)}{L_{i+k}(x_j)}L_{i+k}(x_{j+k})\text{-}L(x_{j+k})
\end{vmatrix} =
$$

$$
\begin{vmatrix}
1 & 1 & \cdots & 1 \\
0 & \dfrac{L(x_j)}{L_i(x_j)}L_i & \cdots & \dfrac{L(x_j)}{L_{i+k}(x_j)}L_{i+k} \\
L(x_{j+1}) & \dfrac{L(x_j)}{L_i(x_j)}L_i(x_{j+1}) & \cdots & \dfrac{L(x_j)}{L_{i+k}(x_j)}L_{i+k}(x_{j+1}) \\
\cdots & \cdots & \cdots & \cdots \\
L(x_{j+k}) & \dfrac{L(x_j)}{L_i(x_j)}L_i(x_{j+k}) & \cdots & \dfrac{L(x_j)}{L_{i+k}(x_j)}L_{i+k}(x_{j+k})
\end{vmatrix} =
$$

$$
\begin{vmatrix}
L(x_j) & L_i(x_j) & \cdots & L_{i+k}(x_j) \\
0 & L_i & \cdots & L_{i+k} \\
L(x_{j+1}) & L_i(x_{j+1}) & \cdots & L_{i+k}(x_{j+1}) \\
\cdots & \cdots & \cdots & \cdots \\
L(x_{j+k}) & L_i(x_{j+k}) & \cdots & L_{i+k}(x_{j+k})
\end{vmatrix}
\dfrac{[L(x_j)]^k}{L_i(x_j)...L_{i+k}(x_j)} \quad .
$$

For the denominator we have

$$
\begin{vmatrix}
1 & \cdots & 1 \\
\dfrac{L(x_j)}{L_i(x_j)}L_i(x_{j+1})\text{-}L(x_{j+1}) & \cdots & \dfrac{L(x_j)}{L_{i+k}(x_j)}L_{i+k}(x_{j+1})\text{-}L(x_{j+1}) \\
\cdots & \cdots & \cdots \\
\dfrac{L(x_j)}{L_i(x_j)}L_i(x_{j+k})\text{-}L(x_{j+k}) & \cdots & \dfrac{L(x_j)}{L_{i+k}(x_j)}L_{i+k}(x_{j+k})\text{-}L(x_{j+k})
\end{vmatrix} =
$$

$$
\begin{vmatrix}
1 & \cdots & 1 \\
\dfrac{L(x_j)}{L_i(x_j)} L_i(x_{j+1}) & \cdots & \dfrac{L(x_j)}{L_{i+k}(x_j)} L_{i+k}(x_{j+1}) \\
\cdots & \cdots & \cdots \\
\dfrac{L(x_j)}{L_i(x_j)} L_i(x_{j+k}) & \cdots & \dfrac{L(x_j)}{L_{i+k}(x_j)} L_{i+k}(x_{j+k})
\end{vmatrix} =
$$

$$
\begin{vmatrix}
L_i(x_j) & \cdots & L_{i+k}(x_j) \\
\cdots & \cdots & \cdots \\
L_i(x_{j+k}) & \cdots & L_{i+k}(x_{j+k})
\end{vmatrix}
\dfrac{[L(x_j)]^k}{L_i(x_j)\ldots L_{i+k}(x_j)} .
$$

Thus the new determinantal identity has been proved since, in order to solve recursively the general interpolation problem in E^*, it must be assumed that $L(x_j)$, $L_i(x_j),\ldots,L_{i+k}(x_j)$ are all different from zero. Consequently, $M_k^{(i,j)}$ can be recursively computed by the H-algorithm with the initalizations

$$
H_o^{(i)} = M_o^{(i,j)} \quad \text{and} \quad g_p(i) = f_{j+p}^{(i,j)} \quad \text{for j fixed}
$$

and we obtain

$$
H_k^{(i)} = M_k^{(i,j)}.
$$

We also have for the numerator of $g_{k-1,k}^{(i)}$

$$
\begin{vmatrix}
g_k(i) & \cdots & g_k(i+k-1) \\
g_1(i) & \cdots & g_1(i+k-1) \\
\cdots & \cdots & \cdots \\
g_{k-1}(i) & \cdots & g_{k-1}(i+k-1)
\end{vmatrix}
= (-1)^{k-1}
\begin{vmatrix}
g_1(i) & \cdots & g_1(i+k-1) \\
\cdots & \cdots & \cdots \\
g_k(i) & \cdots & g_k(i+k-1)
\end{vmatrix}
$$

$$
= (-1)^{k-1}
\begin{vmatrix}
f_{j+1}^{(i,j)} & \cdots & f_{j+1}^{(i+k-1,j)} \\
\cdots & \cdots & \cdots \\
f_{j+k}^{(i,j)} & \cdots & f_{j+k}^{(i+k-1,j)}
\end{vmatrix} =
$$

$$\begin{vmatrix} 1 & 1 & \cdots & 1 \\ 0 & \dfrac{L(x_j)}{L_i(x_j)} L_i(x_{j+1})-L(x_{j+1}) & \cdots & \dfrac{L(x_j)}{L_{i+k-1}(x_j)} L_{i+k-1}(x_{j+1})-L(x_{j+1}) \\ \cdots & \cdots & \cdots & \cdots \\ 0 & \dfrac{L(x_j)}{L_i(x_j)} L_i(x_{j+k})-L(x_{j+k}) & \cdots & \dfrac{L(x_j)}{L_{i+k-1}(x_j)} L_{i+k-1}(x_{j+k})-L(x_{j+k}) \end{vmatrix} (-1)^{k-1} =$$

$$\begin{vmatrix} 1 & 1 & \cdots & 1 \\ L(x_{j+1}) & \dfrac{L(x_j)}{L_i(x_j)} L_i(x_{j+1}) & \cdots & \dfrac{L(x_j)}{L_{i+k-1}(x_j)} L_{i+k-1}(x_{j+1}) \\ \cdots & \cdots & \cdots & \cdots \\ L(x_{j+k}) & \dfrac{L(x_j)}{L_i(x_j)} L_i(x_{j+k}) & \cdots & \dfrac{L(x_j)}{L_{i+k-1}(x_j)} L_{i+k-1}(x_{j+k}) \end{vmatrix} (-1)^{k-1} =$$

$$\begin{vmatrix} L(x_j) & L_i(x_j) & \cdots & L_{i+k-1}(x_j) \\ \cdots & \cdots & \cdots & \cdots \\ L(x_{j+k}) & L_i(x_{j+k}) & \cdots & L_{i+k-1}(x_{j+k}) \end{vmatrix} (-1)^{k-1} \dfrac{[L(x_j)]^{k-1}}{L_i(x_j)...L_{i+k-1}(x_j)} \cdot$$

Thus

$$g_{k-1,k}^{(i)} =$$

$$\begin{vmatrix} L(x_j) & L_i(x_j) & \cdots & L_{i+k-1}(x_j) \\ \cdots & \cdots & \cdots & \cdots \\ L(x_{j+k}) & L_i(x_{j+k}) & \cdots & L_{i+k-1}(x_{j+k}) \end{vmatrix} (-1)^{k-1} \Bigg/ \begin{vmatrix} L_i(x_j) & \cdots & L_{i+k-1}(x_j) \\ \cdots & \cdots & \cdots \\ L_i(x_{j+k-1}) & \cdots & L_{i+k-1}(x_{j+k-1}) \end{vmatrix}$$

$$= -L(x_k^{(i,j)})$$

and the H-algorithm reduces to formula F_{28} of section 4.1. Applying the H-algorithm to f leads to F_{17} again.

We have

$$R_k^{(i,j)} = \begin{vmatrix} L_i(f) & L_i(x_j) & \dots & L_i(x_{j+k}) \\ 0 & x_j & \dots & x_{j+k} \\ L_{i+1}(f) & L_{i+1}(x_j) & \dots & L_{i+1}(x_{j+k}) \\ \dots & \dots & \dots & \dots \\ L_{i+k}(f) & L_{i+k}(x_j) & \dots & L_{i+k}(x_{j+k}) \end{vmatrix} \Big/ \begin{vmatrix} L_i(x_j) & \dots & L_i(x_{j+k}) \\ \dots & \dots & \dots \\ L_{i+k}(x_j) & \dots & L_{i+k}(x_{j+k}) \end{vmatrix}.$$

We shall now prove that we also have

$$R_k^{(i,j)} = \begin{vmatrix} R_0^{(i,j)} & \dots & R_0^{(i,j+k)} \\ e_{i+1}^{(i,j)} & \dots & e_{i+1}^{(i,j+k)} \\ \dots & \dots & \dots \\ e_{i+k}^{(i,j)} & \dots & e_{i+k}^{(i,j+k)} \end{vmatrix} \Big/ \begin{vmatrix} 1 & \dots & 1 \\ e_{i+1}^{(i,j)} & \dots & e_{i+1}^{(i,j+k)} \\ \dots & \dots & \dots \\ e_{i+k}^{(i,j)} & \dots & e_{i+k}^{(i,j+k)} \end{vmatrix}$$

with $R_0^{(i,j)} = \dfrac{L_i(f)}{L_i(x_j)} x_j$ and $e_p^{(i,j)} = \dfrac{L_i(f)}{L_i(x_j)} L_p(x_j) - L_p(f)$.

Thus $e_p^{(i,j)} = L_p(R_0^{(i,j)}) - L_p(f)$ and the above numerator can be written as

$$\begin{vmatrix} 1 & 1 & \dots & 1 \\ 0 & R_0^{(i,j)} & \dots & R_0^{(i,j+k)} \\ 0 & L_{i+1}(R_0^{(i,j)})-L_{i+1}(f) & \dots & L_{i+1}(R_0^{(i,j+k)})-L_{i+1}(f) \\ \dots & \dots & \dots & \dots \\ 0 & L_{i+k}(R_0^{(i,j)})-L_{i+k}(f) & \dots & L_{i+k}(R_0^{(i,j+k)})-L_{i+k}(f) \end{vmatrix} =$$

$$\begin{vmatrix} 1 & 1 & \dots & 1 \\ 0 & R_0^{(i,j)} & \dots & R_0^{(i,j+k)} \\ L_{i+1}(f) & L_{i+1}(R_0^{(i,j)}) & \dots & L_{i+1}(R_0^{(i,j+k)}) \\ \dots & \dots & \dots & \dots \\ L_{i+k}(f) & L_{i+k}(R_0^{(i,j)}) & \dots & L_{i+k}(R_0^{(i,j+k)}) \end{vmatrix} =$$

$$\begin{vmatrix} L_i(f) & L_i(x_j) & ... & L_i(x_{j+k}) \\ 0 & x_j & ... & x_{j+k} \\ L_{i+1}(f) & L_{i+1}(x_j) & ... & L_{i+1}(x_{j+k}) \\ ... & ... & ... & ... \\ L_{i+k}(f) & L_{i+k}(x_j) & ... & L_{i+k}(x_{j+k}) \end{vmatrix} \frac{[L_i(f)]^k}{L_i(x_j)...L_i(x_{j+k})} .$$

For the denominator we have

$$\begin{vmatrix} 1 & ... & 1 \\ L_{i+1}(R_0^{(i,j)})-L_{i+1}(f) & ... & L_{i+1}(R_0^{(i,j+k)})-L_{i+1}(f) \\ ... & ... & ... \\ L_{i+k}(R_0^{(i,j)})-L_{i+k}(f) & ... & L_{i+k}(R_0^{(i,j+k)})-L_{i+k}(f) \end{vmatrix} =$$

$$\begin{vmatrix} 1 & ... & 1 \\ L_{i+1}(R_0^{(i,j)}) & ... & L_{i+1}(R_0^{(i,j+k)}) \\ ... & ... & ... \\ L_{i+k}(R_0^{(i,j)}) & ... & L_{i+k}(R_0^{(i,j+k)}) \end{vmatrix} =$$

$$\begin{vmatrix} L_i(x_j) & ... & L_i(x_{j+k}) \\ ... & ... & ... \\ L_{i+k}(x_j) & ... & L_{i+k}(x_{j+k}) \end{vmatrix} \frac{[L_i(f)]^k}{L_i(x_j)...L_i(x_{j+k})} .$$

Thus the new determinantal formula is proved since, in order to solve recursively the general interpolation problem in E, it is assumed that $L_i(f)$, $L_i(x_j)$,...,$L_i(x_{j+k})$ are all different from zero. $R_k^{(i,j)}$ can be recursively computed by the H-algorithm with the initializations

$$H_0^{(j)} = R_0^{(i,j)} \text{ and } g_p(j) = e_{i+p}^{(i,j)} \text{ for i fixed}$$

and we obtain

$$H_k^{(j)} = R_k^{(i,j)}.$$

We also have for the numerator of $g_{k-1,k}^{(j)}$

$$
\begin{vmatrix}
g_k(j) & \dots & g_k(j+k-1) \\
g_1(j) & \dots & g_1(j+k-1) \\
\dots & \dots & \dots \\
g_{k-1}(j) & \dots & g_{k-1}(j+k-1)
\end{vmatrix}
= (-1)^{k-1}
\begin{vmatrix}
e_{i+1}^{(i,j)} & \dots & e_{i+1}^{(i,j+k-1)} \\
\dots & \dots & \dots \\
e_{i+k}^{(i,j)} & \dots & e_{i+k}^{(i,j+k-1)}
\end{vmatrix}
=
$$

$$
(-1)^{k-1}
\begin{vmatrix}
1 & 1 & \dots & 1 \\
0 & L_{i+1}(R_o^{(i,j)})-L_{i+1}(f) & \dots & L_{i+1}(R_o^{(i,j+k-1)})-L_{i+1}(f) \\
\dots & \dots & \dots & \dots \\
0 & L_{i+k}(R_o^{(i,j)})-L_{i+k}(f) & \dots & L_{i+k}(R_o^{(i,j+k-1)})-L_{i+k}(f)
\end{vmatrix}
=
$$

$$
(-1)^{k-1}
\begin{vmatrix}
1 & 1 & \dots & 1 \\
L_{i+1}(f) & L_{i+1}(R_o^{(i,j)}) & \dots & L_{i+1}(R_o^{(i,j+k-1)}) \\
\dots & \dots & \dots & \dots \\
L_{i+k}(f) & L_{i+k}(R_o^{(i,j)}) & \dots & L_{i+k}(R_o^{(i,j+k-1)})
\end{vmatrix}
=
$$

$$
(-1)^{k-1}
\begin{vmatrix}
L_i(f) & L_i(x_j) & \dots & L_i(x_{j+k-1}) \\
\dots & \dots & \dots & \dots \\
L_{i+k}(f) & L_{i+k}(x_j) & \dots & L_{i+k}(x_{j+k-1})
\end{vmatrix}
\frac{[L_i(f)]^{k-1}}{L_i(x_j)\dots L_i(x_{j+k-1})} .
$$

Thus

$$
g_{k-1,k}^{(j)} =
$$

$$
(-1)^{k-1}
\begin{vmatrix}
L_i(f) & L_i(x_j) & \dots & L_i(x_{j+k-1}) \\
\dots & \dots & \dots & \dots \\
L_{i+k}(f) & L_{i+k}(x_j) & \dots & L_{i+k}(x_{j+k-1})
\end{vmatrix}
\Bigg/
\begin{vmatrix}
L_i(x_j) & \dots & L_i(x_{j+k-1}) \\
\dots & \dots & \dots \\
L_{i+k-1}(x_j) & \dots & L_{i+k-1}(x_{j+k-1})
\end{vmatrix}
$$

$$
= - L_k^{(i,j)}(f) \frac{D_{k+1}^{(i,j)}}{D_k^{(i,j)}}
$$

and the H-algorithm becomes

$$R_k^{(i,j)} = \frac{L_k^{(i,j+1)}(f)\,\dfrac{D_{k+1}^{(i,j+1)}}{D_k^{(i,j+1)}}\,R_{k-1}^{(i,j)} - L_k^{(i,j)}(f)\,\dfrac{D_{k+1}^{(i,j)}}{D_k^{(i,j)}}\,R_{k-1}^{(i,j+1)}}{L_k^{(i,j+1)}(f)\,\dfrac{D_{k+1}^{(i,j+1)}}{D_k^{(i,j+1)}} - L_k^{(i,j)}(f)\,\dfrac{D_{k+1}^{(i,j)}}{D_k^{(i,j)}}}.$$

But $L_{i+k}(x_k^{(i,j)}) = D_{k+1}^{(i,j)}/D_k^{(i,j)}$ and the H-algorithm reduces to F_{29} of section 4.2. Applying the functional L to this relation leads to F_{18} again.

Let us mention that an algorithm more economical that the H-algorithm for the recursive computation of the $H_k^{(n)}$'s is given in [72]. This algorithm can also be used to compute recursively ratios of determinants similar to those of the H_k^n's where the $g_p(n)$'s are replaced by $g_{m+p}(n)$.

Now, in the H-algorithm, let us set

$$e_n = L_n \quad \text{and} \quad g_i(n) = L_n(x_{j+i-1}) \quad \text{for j fixed.}$$

Then

$$g_{k,k+1}^{(n)} = H_k^{(n)}(x_{j+k})$$

$$L_k^{(n,j)} = H_k^{(n)}/g_{k,k+1}^{(n)}$$

and the rule of the H-algorithm becomes

$$g_{k,k+1}^{(n)}\,L_k^{(n,j)} = \frac{L_{k-1}^{(n+1,j)} - L_{k-1}^{(n,j)}}{1/g_{k-1,k}^{(n+1)} - 1/g_{k-1,k}^{(n)}}\,.$$

Applying $H_k^{(n)}$ to x_{j+k} yields

$$H_k^{(n)}(x_{j+k}) = g_{k,k+1}^{(n)} = \frac{L_{k-1}^{(n+1,j)}(x_{j+k}) - L_{k-1}^{(n,j)}(x_{j+k})}{1/g_{k-1,k}^{(n+1)} - 1/g_{k-1,k}^{(n)}}\,.$$

Replacing $g_{k,k-1}^{(n)}$ by this expression in the first relation leads to F_9 of section 4.1.

Now, in the H-algorithm, let us set

$$e_n = x_n \quad \text{and} \quad g_j(n) = L_{i+j-1}(x_n) \quad \text{for i fixed.}$$

Setting

$$D_k^{(n)} = \begin{vmatrix} 1 & \cdots & 1 \\ L_i(x_n) & \cdots & L_i(x_{n+k-1}) \\ \cdots & \cdots & \cdots \\ L_{i+k-2}(x_n) & \cdots & L_{i+k-2}(x_{n+k-1}) \end{vmatrix}$$

we have

$$D_k^{(i,n)} = (-1)^{k-1} \, g_{k-1,k}^{(n)} D_k^{(n)}$$

$$N_{k+1}^{(i,n)} = (-1)^k \, H_k^{(n)} D_{k+1}^{(n)}$$

and

$$x_k^{(i,n)} = - \frac{H_k^{(n)}}{g_{k-1,k}^{(n)}} \frac{D_{k+1}^{(n)}}{D_k^{(n)}} \ .$$

The rule of the H-algorithm becomes

$$x_k^{(i,n)} =$$

$$\frac{D_{k+1}^{(n)} \ g_{k-1,k}^{(n+1)} \ g_{k-2,k-1}^{(n)} x_{k-1}^{(i,n)} \ D_{k-1}^{(n)}/D_k^{(n)} \ - \ g_{k-1,k}^{(n)} \ g_{k-2,k-1}^{(n+1)} x_{k-1}^{(i,n+1)} \ D_{k-1}^{(n+1)}/D_k^{(n+1)}}{D_k^{(n)} \ g_{k-1,k}^{(n)} \qquad\qquad\qquad g_{k-1,k}^{(n+1)} \ - \ g_{k-1,k}^{(n)}} \ .$$

But, by Sylvester's identity, we have

$$D_{k+1}^{(n)} \, D_{k-1}^{(i,n+1)} = D_k^{(n)} \, D_k^{(i,n+1)} - D_k^{(n+1)} \, D_k^{(i,n)}$$

or $\qquad \dfrac{D_{k-1}^{(n+1)} \, D_{k+1}^{(n)}}{D_k^{(n)} \, D_k^{(n+1)}} \ g_{k-2,k-1}^{(n+1)} = g_{k-1,k}^{(n)} - g_{k-1,k}^{(n+1)} \ .$

On the other hand, we also have

$$g_{k,k+1}^{(n)} = L_{i+k}(H_k^{(n)}) = -L_{i+k}(x_k^{(i,n)}) \ g_{k-1,k}^{(n)} \ \frac{D_k^{(n)}}{D_{k+1}^{(n)}} \ .$$

Making use of these two relations, the rule of the H-algorithm finally becomes

$$x_k^{(i,n)} = x_{k-1}^{(i,n+1)} - \frac{L_{i+k-1}(x_{k-1}^{(i,n+1)})}{L_{i+k-1}(x_{k-1}^{(i,n)})} \ x_{k-1}^{(i,n)}$$

which is F_5 of section 4.1.

5- APPLICATIONS

5.1 - Sequence transformations

The ratios of determinants studied in the preceding sections and the recursive algorithms for their computation have applications in the general interpolation problem and in sequence transformations (which are extrapolation methods) used in numerical analysis to accelerate the convergence. Such sequence transformations, which can also be considered as projection methods, can be used to construct fixed point iterations, an ancient approach which has recently received much attention [114, 166, 167, 180, 181, 183]. As we shall see now all these methods and algorithms are direct applications of the results given in the previous sections.

Let us begin by interpolation since it was our first objective. The Neville-Aitken scheme is a well known procedure for the recursive computation of interpolation polynomials. Instead of interpolating by a polynomial one may want to interpolate by a linear combination of functions forming a complete Chebyshev system and try to find the corresponding generalization of the Neville-Aitken scheme. This was done by Mühlbach in a series of papers. He first gave a generalization of divided differences [139] which is now in fact a particular case of that given in section 4.1. In 1976, Mühlbach [140] obtained a recursive scheme for computing the $P_k^{(n)}(x)$'s given by

$$P_k^{(n)}(x) = a_0 g_0(x) + \dots + a_k g_k(x)$$

such that $P_k^{(n)}(t_i) = w_i$ for $i = n,...,n+k$. This algorithm, called the MNA-algorithm (for Mühlbach-Neville-Aitken) is in fact the H-algorithm with the initializations

$$H_0^{(n)} = w_n \, g_0(x)/g_0(t_n) \qquad g_{0,i}^{(n)} = g_i(t_n)g_0(x)/g_0(t_n)-g_i(x)$$

and we obtain

$$H_k^{(n)} = P_k^{(n)}(x).$$

If we set $L_n(x_j) = g_j(t_n)$, $L_n(f) = f(t_n) = w_n$ and if L is defined by $L(x_j)=g_j(x)$ where x is fixed then we also have

$$P_k^{(n)}(x) = M_k^{(n,o)}(f).$$

The other algorithms given in section 4.1 provide other recursive methods for computing the $P_k^{(n)}(x)$'s.

The MNA-algorithm can be adapted to treat the general interpolation problem, see [20] where a subroutine is also given. Various proofs of the MNA-algorithm can be found in the literature [19, 88, 141]. Since Sylvester's identity is related to Gaussian elimination, to Schur complements and to the bordering method, the development of the subject gave rise to several papers on these connections and on some extensions [74, 76, 77, 144, 148]. The algorithm was extended to interpolation by rational functions [89, 90, 91, 93, 126] (see also [125]), to quadratic approximation [124, 127], to vector orthogonal polynomials [103] and to multivariate interpolation [51, 53, 55, 75, 146, 184]. More comments and references on these topics can be found in [33].

Let now (S_n) be a sequence of elements of E. We consider the general interpolation problem of finding

$$R_k^{(o,n)} = a_0 S_n + ... + a_k S_{n+k}$$

such that $L_0(R_k^{(o,n)}) = 1$ and $L_i(R_k^{(o,n)}) = 0$ for $i = 1,...,k$.

We set $L_i(S_j) = g_i(j)$ and we assume that $\forall j, g_0(j) = 1$. Thus we have

$$
R_k^{(o,n)} = - \begin{vmatrix} 0 & S_n & \dots & S_{n+k} \\ 1 & 1 & \dots & 1 \\ 0 & g_1(n) & \dots & g_1(n+k) \\ \dots & \dots & \dots & \dots \\ 0 & g_k(n) & \dots & g_k(n+k) \end{vmatrix} \Big/ \begin{vmatrix} 1 & \dots & 1 \\ g_1(n) & \dots & g_1(n+k) \\ \dots & \dots & \dots \\ g_k(n) & \dots & g_k(n+k) \end{vmatrix}
$$

$$
= \begin{vmatrix} S_n & \dots & S_{n+k} \\ g_1(n) & \dots & g_1(n+k) \\ \dots & \dots & \dots \\ g_k(n) & \dots & g_k(n+k) \end{vmatrix} \Big/ \begin{vmatrix} 1 & \dots & 1 \\ g_1(n) & \dots & g_1(n+k) \\ \dots & \dots & \dots \\ g_k(n) & \dots & g_k(n+k) \end{vmatrix} .
$$

Thus these ratios of determinants can be recursively computed via the H-algorithm and we have $H_k^{(n)} = R_k^{(o,n)}$.

When (S_n) is a scalar sequence, the preceding ratio of determinants includes most of the sequence transformations used for accelerating the convergence of (S_n) and thus the H-algorithm (called the E-algorithm in this particular case) provides a recursive method for computing the numbers $H_k^{(n)}$ (denoted by $E_k^{(n)}$ in this case) without computing the determinants involved in their definition. This algorithm obtained independently by several authors [18, 87, 133, 168] is the most general extrapolation algorithm actually known. It can be used for the implementation of several famous transformations such as that of Shanks [170] which was usually implemented by the ε-algorithm of Wynn [189], rational extrapolation for which the ρ-algorithm was used [190] and many others [13]. Division by zero or numerical instability in the algorithm can be avoided (at least partly) by using the multistep formula given in section 4.2 for the H-algorithm. A subroutine for the E-algorithm can be found in [20] while other subroutines are given in [14] (see also [39]). Considerations on the numerical stability of convergence acceleration methods are discussed in [49].

The E-algorithm can also be recovered by solving the general interpolation problem in E*. Let L_n be the linear functional on the space of sequences $S = (S_n)$ such that $L_n(S) = S_n$. We consider the problem of finding

$$
M_k^{(n,o)} = b_0 L_n + \dots + b_k L_{n+k}
$$

such that $M_k^{(n,o)}(g_0) = 1$ and $M_k^{(n,o)}(g_i) = 0$ for $i = 1,\dots,k$, where the g_i's are sequences.

We assume that $\forall j$, $g_0(j) = 1$. Thus we have

$$
M_k^{(n,o)} = \left| \begin{array}{ccc} L_n & \dots & L_{n+k} \\ g_1(n) & \dots & g_1(n+k) \\ \dots & \dots & \dots \\ g_k(n) & \dots & g_k(n+k) \end{array} \right| \Bigg/ \left| \begin{array}{ccc} 1 & \dots & 1 \\ g_1(n) & \dots & g_1(n+k) \\ \dots & \dots & \dots \\ g_k(n) & \dots & g_k(n+k) \end{array} \right| .
$$

$M_k^{(n,o)}$ is the so-called extrapolation operator which can be recursively computed by the H-algorithm and we have

$$
M_k^{(n,o)} (S) = E_k^{(n)} .
$$

The various approaches leading to the E-algorithm have been reviewed in [33] where many references connected with it are to be found.

In the E-algorithm we have

$$
E_k^{(n)} = \left| \begin{array}{ccc} S_n & \dots & S_{n+k} \\ g_1(n) & \dots & g_1(n+k) \\ \dots & \dots & \dots \\ g_k(n) & \dots & g_k(n+k) \end{array} \right| \Bigg/ \left| \begin{array}{ccc} \Delta g_1(n) & \dots & \Delta g_1(n+k-1) \\ \dots & \dots & \dots \\ \Delta g_k(n) & \dots & \Delta g_k(n+k-1) \end{array} \right| .
$$

Let us set $E_k^{(n)} = e_{2k}^{(n)}$ and introduce the intermediate quantities

$$
e_{2k+1}^{(n)} = \left| \begin{array}{ccc} \Delta \alpha_n & \dots & \Delta \alpha_{n+k} \\ \Delta g_1(n) & \dots & \Delta g_1(n+k) \\ \dots & \dots & \dots \\ \Delta g_k(n) & \dots & \Delta g_k(n+k) \end{array} \right| \Bigg/ \left| \begin{array}{ccc} \Delta S_n & \dots & \Delta S_{n+k} \\ \Delta g_1(n) & \dots & \Delta g_1(n+k) \\ \dots & \dots & \dots \\ \Delta g_k(n) & \dots & \Delta g_k(n+k) \end{array} \right|
$$

where (α_n) is an arbitrary given sequence. We set

$$\mu_k^{(n)} = \frac{\begin{vmatrix} \Delta\alpha_n & \cdots & \Delta\alpha_{n+k} \\ \Delta S_n & \cdots & \Delta S_{n+k} \\ \Delta g_1(n) & \cdots & \Delta g_1(n+k) \\ \cdots & \cdots & \cdots \\ \Delta g_{k-1}(n) & \cdots & \Delta g_{k-1}(n+k) \end{vmatrix} \cdot \begin{vmatrix} g_1(n+1) & \cdots & g_1(n+k) \\ \cdots & \cdots & \cdots \\ g_k(n+1) & \cdots & g_k(n+k) \end{vmatrix}}{\begin{vmatrix} \Delta S_{n+1} & \cdots & \Delta S_{n+k} \\ \Delta g_1(n+1) & \cdots & \Delta g_1(n+k) \\ \cdots & \cdots & \cdots \\ \Delta g_{k-1}(n+1) & \cdots & \Delta g_{k-1}(n+k) \end{vmatrix} \cdot \begin{vmatrix} \Delta g_1(n) & \cdots & \Delta g_1(n+k-1) \\ \cdots & \cdots & \cdots \\ \Delta g_k(n) & \cdots & \Delta g_k(n+k-1) \end{vmatrix}} .$$

It was proved in [46] that

$$e_{2k}^{(n)} = e_{2k-2}^{(n+1)} + \frac{\mu_k^{(n)}}{e_{2k-1}^{(n+1)} - e_{2k-1}^{(n)}}$$

$$e_{2k+1}^{(n)} = e_{2k-1}^{(n+1)} + \frac{\mu_k^{(n)}}{e_{2k}^{(n+1)} - e_{2k}^{(n)}}$$

with $e_{-2}^{(n)} = e_{-1}^{(n)} = 0$ and $e_0^{(n)} = S_n$.

Clearly this algorithm is a generalization of Wynn's ε-algorithm [189] and of its first generalisation [11] but it does not reduce to other rhombus algorithms.

For the moment, no recursive scheme for computing the coefficients $\mu_k^{(n)}$ is known. Thus special parameters (α_n) have to be chosen so that the determinants in $\mu_k^{(n)}$ can be easily computed.

As in the ε-algorithm, the $e_{2k+1}^{(n)}$ are intermediate results without any meaning and they can be eliminated thus leading to a generalization to the E-algorithm of Wynn's cross rule for the ε-algorithm [191]

$$\frac{\mu_k^{(n)}}{e_{2k}^{(n)} - e_{2k}^{(n+1)}} + \frac{\mu_k^{(n+1)}}{e_{2k}^{(n+2)} - e_{2k}^{(n+1)}} = \frac{\mu_k^{(n+1)}}{e_{2k-2}^{(n+2)} - e_{2k}^{(n+1)}} + \frac{\mu_{k+1}^{(n)}}{e_{2k+2}^{(n)} - e_{2k}^{(n+1)}} .$$

Thus this generalized ϵ-algorithm fits into the rhombus rules examined by Cordellier [48] but it does not possess in general the homographic invariance property since in general $\mu_k^{(n)} \neq \mu_{k+1}^{(n)}$ which is a necessary and sufficient condition for it. An algorithm possessing this property can be transformed, when necessary, into a more reliable and stable one. The ϵ-algorithm and its first generalization have the homographic invariance property.

The E-algorithm (when (S_n) is a scalar sequence) or, more generally the H-algorithm (that is the previous $R_k^{(o,n)}$) have been studied and they received many applications. On a general theory of extrapolation methods and the algorithmic aspects see [2, 29, 31, 71, 92, 142, 143, 144, 145]. The E-algorithm have applications to multivariate Padé approximation [52], to partial Padé approximation [157] and to the acceleration of double sequences [51, 54, 85] among others. It can be used for the implementation of composite sequence transformations [24], for the solution of systems of linear equations in the least squares sense [20, 138] and for the computation of Stieltjes polynomials [34].

A particular case arises when $g_i(n) = \langle y, \Delta S_{n+i-1} \rangle$ where $y \in E^*$; it is the so-called topological ϵ-algorithm [12] which, in the vector case, is connected to the bi-conjugate gradient method [17, p. 189].

This algorithm has received much attention. It has applications to fixed point methods where it provides quadratic convergence without computing derivatives nor inverting any matrix [40]. Its implementation was studied in [183] and another application is given in [182].

Depending on the choice for the $g_i(n)$'s in the H-algorithm, several projection methods can be obtained such as the topological ϵ-algorithm, the minimal polynomial extrapolation method [45], Henrici's method [98], the $S\beta$-algorithm [113] and the reduced rank extrapolation [66, 135]. These methods, which can be used for solving systems of linear or nonlinear equations and eigenvalues problems, have been recently studied in a unified framework [166, 173, 174, 175, 176, 180, 181].

Let us only give a brief description of the $S\beta$-algorithm obtained by Jbilou [113]. In this algorithm the following ratios of determinants are considered

$$S_k^{(n)} = \begin{vmatrix} S_n & ... & S_{n+k} \\ L_1(\Delta S_n) & ... & L_1(\Delta S_{n+k}) \\ ... & ... & ... \\ L_k(\Delta S_n) & ... & L_k(\Delta S_{n+k}) \end{vmatrix} \Bigg/ \begin{vmatrix} 1 & ... & 1 \\ L_1(\Delta S_n) & ... & L_1(\Delta S_{n+k}) \\ ... & ... & ... \\ L_k(\Delta S_n) & ... & L_k(\Delta S_{n+k}) \end{vmatrix}$$

$$\beta_k^{(n)} = \begin{vmatrix} \Delta S_n & ... & \Delta S_{n+k} \\ L_1(\Delta S_n) & ... & L_1(\Delta S_{n+k}) \\ ... & ... & ... \\ L_k(\Delta S_n) & ... & L_k(\Delta S_{n+k}) \end{vmatrix} \Bigg/ \begin{vmatrix} 1 & ... & 1 \\ L_1(\Delta S_n) & ... & L_1(\Delta S_{n+k}) \\ ... & ... & ... \\ L_k(\Delta S_n) & ... & L_k(\Delta S_{n+k}) \end{vmatrix} .$$

It is proved, by a direct application of Sylvester's identity, that it holds

$$S_k^{(n)} = \frac{L_k(\beta_{k-1}^{(n+1)})S_{k-1}^{(n)} - L_k(\beta_{k-1}^{(n)})S_{k-1}^{(n+1)}}{L_k(\beta_{k-1}^{(n+1)}) - L_k(\beta_{k-1}^{(n)})}$$

$$\beta_k^{(n)} = \frac{L_k(\beta_{k-1}^{(n+1)})\beta_{k-1}^{(n)} - L_k(\beta_{k-1}^{(n)})\beta_{k-1}^{(n+1)}}{L_k(\beta_{k-1}^{(n+1)}) - L_k(\beta_{k-1}^{(n)})}$$

with $\beta_0^{(n)} = \Delta S_n$ and $S_0^{(n)} = S_n$. (On this algorithm, see [114]). Setting $g_{k,i}^{(n)} = L_i(\beta_k^{(n)})$ and applying L_i to this second relation, leads to the H-algorithm again.

In what precedes (and it will also be the case in what follows) $R_k^{(i,j)}$ and $M_k^{(i,j)}$ depend on three indexes while $H_k^{(n)}$ and $E_k^{(n)}$ only depend on two. Thus, using the recursive formulae of section 4.1, we now have possible extensions of the preceding procedures.

Let $f : \mathbf{R}^p \to \mathbf{R}^p$. Fixed point methods for solving $f(x) = 0$ usually produce a sequence of vectors (x_n) such that

$$x_{n+1} = x_n - J_n^{-1} f(x_n)$$

where J_n is either the Jacobian matrix of f at the point x_n (Newton's method) or an approximation of it. We have

$$x_{n+1} = \begin{vmatrix} x_n & I \\ f(x_n) & J_n \end{vmatrix} / |J_n|$$

$$= x_n + \begin{vmatrix} 0 & I \\ f(x_n) & J_n \end{vmatrix} / |J_n| .$$

Let us denote by e_i the ith vector of the canonical basis of $\mathbf{R}^p$ (that is the ith column of the matrix I) and by y_i the ith row of J_n. We also set $L_i(e_j) = (y_i, e_j)$. Thus

$$x_n - x_{n+1} = - \begin{vmatrix} 0 & e_1 & ... & e_p \\ f_1(x_n) & L_1(e_1) & ... & L_1(e_p) \\ ... & ... & ... & ... \\ f_p(x_n) & L_p(e_1) & ... & L_p(e_p) \end{vmatrix} / |J_n|$$

which shows that $x_n - x_{n+1}$ is the solution of the interpolation problem

$$L_i(x_n - x_{n+1}) = f_i(x_n) \qquad i = 1,...,p.$$

For example in the case $p=1$ and $y_1 = f_1'(x_n)$, which corresponds to Newton's method in $\mathbf{R}$, we have

$$L_1(x_n - x_{n+1}) = (x_n - x_{n+1}) f_1'(x_n) = f_1(x_n)$$

that is

$$x_{n+1} = x_n - f_1(x_n) / f_1'(x_n).$$

If $p > 1$, the components of y_i are the partial derivatives of f_i at the point x_n.

The same interpretation also holds for the other fixed point methods already mentioned such as Henrici's or the topological ε-algorithm. The case of the conjugate gradient method was already studied in [17, p. 84-90, 186-189].

In [18] an extension of the E-algorithm to the vector case was given. Let S_n, $g_1(n),...,g_k(n)$ be vectors (or more generaly elements of a vector space E) and y a vector (or more generally an element of E^*). We consider the ratios of determinants

$$
E_k^{(n)} = \cfrac{\begin{vmatrix} S_n & g_1(n) & ... & g_k(n) \\ (y,\Delta S_n) & (y,\Delta g_1(n)) & ... & (y,\Delta g_k(n)) \\ ... & ... & ... & ... \\ (y,\Delta S_{n+k-1}) & (y,\Delta g_1(n+k-1)) & ... & (y,\Delta g_k(n+k-1)) \end{vmatrix}}{\begin{vmatrix} (y,\Delta g_1(n)) & ... & (y,\Delta g_k(n)) \\ ... & ... & ... \\ (y,\Delta g_1(n+k-1)) & ... & (y,\Delta g_k(n+k-1)) \end{vmatrix}}
$$

and a similar ratio for $g_{k,i}^{(n)}$ by replacing the first column in the numerator by $(g_i(n), (y,\Delta g_i(n)),...,(y,\Delta g_i(n+k-1)))^T$. $(.,.)$ denotes the duality product between E and E^*. The following recurrence relations were proved to hold

$$
E_0^{(n)} = S_n \qquad\qquad g_{0,i}^{(n)} = g_i(n) \qquad\qquad n = 0,1,... \ ; \ i = 1,2,...
$$

$$
E_k^{(n)} = E_{k-1}^{(n)} - \frac{(y,\Delta E_{k-1}^{(n)})}{(y,\Delta g_{k-1,k}^{(n)})} g_{k-1,k}^{(n)} \qquad\qquad k = 1,2,... \ ; \ n = 0,1,...
$$

$$
g_{k,i}^{(n)} = g_{k-1,i}^{(n)} - \frac{(y,\Delta g_{k-1,i}^{(n)})}{(y,\Delta g_{k-1,k}^{(n)})} g_{k-1,k}^{(n)} \qquad\qquad k = 1,2,... \ ; \ n = 0,1,... \ ; \ i = k+1,...
$$

where Δ acts on the superscript n.

Using the connection between interpolation and biorthogonality explained in section 3 (just before 3.1) we have

$$
R_{k-1}^{(1,n)} = S_n - E_k^{(n)}
$$

where $R_{k-1}^{(1,n)}$ is given by the same ratio of determinants as $E_k^{(n)}$ where S_n in the first row and first column is replaced by 0 and satisfies the interpolation conditions

$$L_i(R_{k-1}^{(1,n)}) = (y, \Delta S_{n+i-1}) \qquad\qquad i = 1,...,k.$$

Of course a similar interpretation holds for the $g_{k,i}^{(n)}$'s and these ratios of determinants can be put in the framework of the H-algorithm as seen in section 4.2. An extension where y is replaced by a sequence (y_n) is studied in [188].

A transformation which can be considered as intermediate between the H-algorithm and the topological ε-algorithm is the G-transformation [84]. In this transformation we consider the ratios

$$G_k^{(n)} = \begin{vmatrix} S_n & ... & S_{n+k} \\ c_n & .... & c_{n+k} \\ ... & ... & ... \\ c_{n+k-1} & ... & c_{n+2k-1} \end{vmatrix} \Bigg/ \begin{vmatrix} 1 & ... & 1 \\ c_n & ... & c_{n+k} \\ ... & ... & ... \\ c_{n+k-1} & ... & c_{n+2k-1} \end{vmatrix}$$

where the c_n's are scalars and the S_n's elements of a vector space. Setting

$$H_k^{(n)} = \begin{vmatrix} c_n & .... & c_{n+k-1} \\ ... & ... & ... \\ c_{n+k-1} & ... & c_{n+2k-2} \end{vmatrix}$$

$$\bar{H}_k^{(n)} = \begin{vmatrix} \Delta c_n & .... & \Delta c_{n+k-1} \\ ... & ... & ... \\ \Delta c_{n+k-1} & ... & \Delta c_{n+2k-2} \end{vmatrix} = \begin{vmatrix} 1 & ... & 1 \\ c_n & ... & c_{n+k} \\ ... & ... & ... \\ c_{n+k-1} & ... & c_{n+2k-1} \end{vmatrix}$$

$$r_k^{(n)} = H_k^{(n)}/\bar{H}_{k-1}^{(n)} \quad \text{and} \quad s_k^{(n)} = \bar{H}_k^{(n)}/H_k^{(n)}$$

it was proved by Pye and Atchison [160] that

$$(1 - r_{k+1}^{(n+1)}/r_{k+1}^{(n)})G_{k+1}^{(n)} = G_{k+1}^{(n+1)} - G_k^{(n)} r_{k+1}^{(n+1)}/r_{k+1}^{(n)}$$

$$s_{k+1}^{(n+1)}/s_k^{(n)} = 1 + r_{k+1}^{(n)}/r_k^{(n+1)}$$

$$r_{k+1}^{(n+1)}/r_{k+1}^{(n)} = 1 + s_{k+1}^{(n)}/s_k^{(n+1)}$$

with $G_0^{(n)} = S_n$, $s_0^{(n)} = 1$, $r_1^{(n)} = c_n$.

If the H-algorithm is applied with $H_o^{(n)} = S_n$ and $g_i^{(n)} = c_{n+i-1}$ then $r_k^{(n)} = (-1)^{k-1} g_{k-1,k}^{(n)}$, $H_k^{(n)} = G_k^{(n)}$ and the rule of the H-algorithm is identical with that of the G-algorithm. Moreover if Schweins' formula and Sylvester's formula are applied to $\bar{H}_k^{(n)}$ then we directly obtain the preceding recursive formulae for the $r_k^{(n)}$'s and the $s_k^{(n)}$'s. If $c_n = <y, \Delta S_n>$ then the $G_k^{(n)}$'s are identical with the $\varepsilon_{2k}^{(n)}$ given by the topological ε-algorithm but with less arithmetical operations and less storage as shown in [15]. The rs-algorithm is related to the qd-algorithm since

$$q_{k+1}^{(n)} = r_{k+1}^{(n+1)} s_k^{(n+1)} / r_{k+1}^{(n)} s_k^{(n)}$$

$$e_{k+1}^{(n)} = r_{k+2}^{(n)} s_{k+1}^{(n)} / r_{k+1}^{(n+1)} s_k^{(n+1)}.$$

Thus

$$q_{k+1}^{(n)} = g_{k,k+1}^{(n+1)} [1/g_{k,k+1}^{(n)} - 1/g_{k-1,k}^{(n+1)}]$$

$$e_{k+1}^{(n)} = g_{k+1,k+2}^{(n)} [1/g_{k,k+1}^{(n+1)} - 1/g_{k,k+1}^{(n)}].$$

Thanks to the rs-algorithm and to the connection with formal orthogonal polynomials many new relations for Shanks' transformation (that is the scalar ε-algorithm) were given in [15]. Of course these relations also hold for the topological ε-algorithm. They can be deduced from the relations given in sections 4.1 and 4.2.

Finally in [21] some algorithms and ratios of determinants more directly connected with those of the previous sections were given. First we consider the following ratio of determinants

$$E_k = \begin{vmatrix} y & x_1 & \dots & x_k \\ L_1(y) & L_1(x_1) & \dots & L_1(x_k) \\ \dots & \dots & \dots & \dots \\ L_k(y) & L_k(x_1) & \dots & L_k(x_k) \end{vmatrix} \Bigg/ \begin{vmatrix} L_1(x_1) & \dots & L_1(x_k) \\ \dots & \dots & \dots \\ L_k(x_1) & \dots & L_k(x_k) \end{vmatrix}$$

and $g_{k,i}$ obtained by replacing the first column of the numerator by $(x_i, L_1(x_i),\dots,L_k(x_i))^T$. It was proved that we have the following recursive scheme called the RPA (recursive projection algorithm)

$$E_0 = y \ , \ \ g_{0,i} = x_i \ \ , \ \ i \geq 1$$

$$E_k = E_{k-1} - \frac{L_k(E_{k-1})}{L_k(g_{k-1,k})} \, g_{k-1,k} \qquad\qquad k > 0$$

$$g_{k,i} = g_{k-1,i} - \frac{L_k(g_{k-1,i})}{L_k(g_{k-1,k})} \, g_{k-1,k} \qquad\qquad i > k > 0.$$

Of course, due to the connection between interpolation and biorthogonality, we have

$$E_k = y - R_{k-1}^{(1,1)}$$

where $R_{k-1}^{(1,1)} \in$ Span $(x_1, \ldots, x_k)$ satisfies

$$L_i(R_{k-1}^{(1,1)}) = L_i(y) \qquad\qquad i = 1, \ldots, k.$$

We know, from the previous sections, that

$$R_{k-1}^{(1,1)} = R_{k-2}^{(1,1)} + L_{k-1}^{(1,1)} \, (y) \, x_{k-1}^{(1,1)}.$$

Moreover it is easy to see that $x_{k-1}^{(1,1)} = g_{k-1,k}$ and that $L_{k-1}^{(1,1)}(y) = L_k(E_{k-1})/L_k(g_{k-1,k})$. Thus both formulae are the same.

Then the following ratios of determinants are considered

$$e_k^{(i)} = \left| \begin{matrix} x_i & x_{i+1} & \ldots & x_{i+k} \\ L_1(x_i) & L_1(x_{i+1}) & \ldots & L_1(x_{i+k}) \\ \ldots & \ldots & \ldots & \ldots \\ L_k(x_i) & L_k(x_{i+1}) & \ldots & L_k(x_{i+k}) \end{matrix} \right| \Bigg/ \left| \begin{matrix} L_1(x_{i+1}) & \ldots & L_1(x_{i+k}) \\ \ldots & \ldots & \ldots \\ L_k(x_{i+1}) & \ldots & L_k(x_{i+k}) \end{matrix} \right|$$

and it is proved that the more compact recursive scheme (the CRPA where C stands for compact) holds

$$e_0^{(i)} = x_i \qquad\qquad i \geq 0$$

$$e_k^{(i)} = e_{k-1}^{(i)} - \frac{L_k(e_{k-1}^{(i)})}{L_k(e_{k-1}^{(i+1)})} e_{k-1}^{(i+1)} \qquad i \geq 0,\, k \geq 1.$$

We have again

$$e_k^{(i)} = x_i - R_{k-1}^{(1,i+1)}$$

where $R_{k-1}^{(1,i+1)} \in \text{Span}(x_{i+1},\ldots,x_{i+k})$ satisfies

$$L_j(R_{k-1}^{(1,i+1)}) = L_j(x_i) \quad \text{for} \quad j = 1,\ldots,k.$$

It is easy to check that

$$\frac{L_k(e_{k-1}^{(i)})}{L_k(e_{k-1}^{(i+1)})} e_{k-1}^{(i+1)} = (-1)^{k-1} \frac{D_k^{(1,i)}}{D_k^{(1,i+1)}} x_{k-1}^{(1,i+1)}$$

with $D_k^{(1,i)}/D_k^{(1,i+1)} = (-1)^{k-1} L_{k-1}^{(1,i+1)}(x_i)$ and thus we finally obtain the formula of section 4.1

$$R_{k-1}^{(1,i+1)} = R_{k-2}^{(1,i+1)} + L_{k-1}^{(1,i+1)}(x_i) x_{k-1}^{(1,i+1)}.$$

These two algorithms (RPA and CRPA) are related to recursive projection in an inner product space, to Fourier expansion, to Rosen's and Henrici's methods are shown in [21]. They can be used for implementing the E-algorithm or another sequence transformation due to Germain-Bonne [79] or some so-called confluent algorithms [21]. They have been used also to compute recursively the vector Padé approximants of Van Iseghem [101] or in some methods for solving systems of linear equations [121].

The solution of a system of linear equations $Ax = b$ can be considered as an interpolation problem. Let a_i be the ith row of the matrix A and let us define the functionals L_i by

$$L_i(.) = (a_i,.).$$

Then $Ax = b$ is equivalent to

$$L_i(x) = b_i \quad \text{for } i = 1,\ldots,n$$

where b_i is the ith component of the vector b.

The solution of this interpolation problem can be obtain via the RPA as explained in [21, sect. 1] and, thus, a method due to Sloboda [178, 179] is recovered if x_i is the ith vector of the canonical basis of R^n (a null vector x_0 is also needed).

Moreover the solution x of the linear system can be expressed as

$$
x = - \begin{vmatrix} 0 & x_1 & \ldots & \ldots x_n \\ b_1 & a_{11} & \ldots & a_{1n} \\ \ldots & \ldots & \ldots & \ldots \\ b_n & a_{n1} & \ldots & a_{nn} \end{vmatrix} \Bigg/ \begin{vmatrix} a_{11} & \ldots & a_{1n} \\ \ldots & \ldots & \ldots \\ a_{n1} & \ldots & a_{nn} \end{vmatrix} = - \begin{vmatrix} 0 & I \\ b & A \end{vmatrix} \Bigg/ |A| \ .
$$

Compare with the formula for Newton's method given in section 5.1.

As stated in [1], Sloboda's method (and thus the RPA for the solution of a system of linear equations) is a particular case of the so-called ABS method due to Abaffy, Broyden and Spedicato as are also many other terminating algorithms for solving linear systems.

In [192], Wynn considered the ratios of determinants

$$
w_{2k}^{(i)} = H_{k+1}^{(i)} / H_k^{(i+2)}
$$

where $H_k^{(i)} = \begin{vmatrix} x_i & \ldots & x_{i+k-1} \\ \ldots & \ldots & \ldots \\ x_{i+k-1} & \ldots & x_{i+2k-2} \end{vmatrix}$ and the x_i's are numbers and he proved

that they can be computed by the recursive scheme

$$
w_{-2}^{(i)} = \infty \qquad\qquad w_0^{(i)} = x_i
$$

$$
w_{2k+2}^{(i)} = w_{2k}^{(i)} - [w_{2k}^{(i+1)}]^2 [1/w_{2k}^{(i+2)} - 1/w_{2k-2}^{(i+2)}] \ .
$$

If we define, in the CRPA, L_j by $L_j(x_i) = x_{i+j}$ then

$$
e_k^{(i)} = w_{2k}^{(i)} \ .
$$

If we compare the rules of the w-algorithm and of the CRPA we shall have

$$
e_k^{(i+1)} [1/e_k^{(i+2)} - 1/e_{k-1}^{(i+2)}] = L_{k+1}(e_k^{(i)})/L_{k+1}(e_k^{(i+1)})
$$

which is indeed true by Sylvester's formula and since

$$L_{k+1}(e_k^{(i)}) = (-1)^k H_{k+1}^{(i+1)}/H_k^{(i+2)}.$$

The w-algorithm is useful in the computation of a diagonal of the ε-algorithm or in the implementation of the so-called confluent forms of the ε- and ρ-algorithms. On these questions see [15, 39].

In [21] it was proved that, if we set

$$\bar{e}_k^{(i)} = (-1)^k N_{k+1}^{(1,i)}/D_k^{(1,i)}$$

then

$$\bar{e}_k^{(i)} = \frac{L_k(\bar{e}_{k-1}^{(i+1)})}{L_k(\bar{e}_{k-1}^{(i)})} \bar{e}_{k-1}^{(i)} - \bar{e}_{k-1}^{(i+1)}$$

with $\bar{e}_0^{(i)} = x_i$, which is exactly F_5.

Finally, in the same paper, the following ratios were considered

$$\tilde{e}_k^{(i)} = \begin{vmatrix} x_i & \dots & x_{i+k} \\ L_1(x_i) & \dots & L_1(x_{i+k}) \\ \dots & \dots & \dots \\ L_k(x_i) & \dots & L_k(x_{i+k}) \end{vmatrix} \Bigg/ \begin{vmatrix} 1 & \dots & 1 \\ L_1(x_i) & \dots & L_1(x_{i+k}) \\ \dots & \dots & \dots \\ L_k(x_i) & \dots & L_k(x_{i+k}) \end{vmatrix}$$

which are exactly those considered in the H-algorithm.
It was proved that

$$\tilde{e}_k^{(i)} = \frac{L_k(\tilde{e}_{k-1}^{(i+1)})\tilde{e}_{k-1}^{(i)} - L_k(\tilde{e}_{k-1}^{(i)})\tilde{e}_{k-1}^{(i+1)}}{L_k(\tilde{e}_{k-1}^{(i+1)}) - L_k(\tilde{e}_{k-1}^{(i)})}$$

with $\tilde{e}_0^{(i)} = x_i$.

Since $L_k(\tilde{e}_{k-1}^{(i)}) = g_{k-1,k}^{(i)}$ if we set $g_j(i) = L_j(x_i)$, the preceding algorithm is the H-algorithm and it can be used for implementing Henrici's sequence transformation [98].

Thus almost all the sequence transformations and the corresponding algorithms which are known fit into our framework and are particular cases of the results given in the preceding sections. Moreover the algorithms of sections 4.1 and 4.2 provide other possible recursive schemes for the implementation of these transformations and are thus useful in convergence acceleration, orthogonal polynomials, Padé approximation and fixed point methods. In particular the following algorithms have or can be studied in our context : E-algorithm, H-algorithm and Henrici's transformation, RPA and CRPA, composite sequence transformations for scalar and vector sequences, the secant method and its various possible generalizations [166], the method of Pugachev [159], the conjugate and bi-conjugate gradient methods, the generalized conjugate residual method [67], the method of Arnoldi [165], the topological ε-algorithm and its variants [12], the method of Germain-Bonne [79, property 12, p. 17], the minimal polynomial extrapolation [45], the reduced rank extrapolation [66, 159], the generalized minimal polynomial extrapolation [79], the generalization of Wimp of the topological E-algorithm [188]. For a review and theoretical results on these methods see [40, 114, 166, 167, 180, 181]. Least squares extrapolation as described in [20] and [49] can also be put into this framework as well as rational interpolation [56].A quite complete exposition can be found in [39] which also contains subroutines.

5.2 - Linear multistep methods.

We consider the differential equation

$$y'(x) = f(x, y(x)).$$

Let us define the following operators

$$Dg(x) = g'(x)$$
$$Eg(x) = g(x+h)$$
$$\Delta g(x) = (E-I)\, g(x) = g(x+h) - g(x)$$

where h is a positive parameter (the step size).
It is well known that formally [99]

$$E = e^{hD}$$

or

$$D = \frac{1}{h} \, Log \, (I+\Delta)$$

an identity first given by George Boole in 1859 in his *Treatise on Differential Equations.*

Let $R(t) = A(t)/B(t)$ be a rational approximation to Log $(1+t)$. Then the differential equations $Dy(x) = \frac{1}{h}$ Log $(I+\Delta)$ $y(x) = f(x, y(x))$ can be replaced by the approximate equation

$$A(\Delta)y_n = hB(\Delta) f_n$$

where y_n is an approximation of the solution y at the point $x_n = x_0 + nh$ and $f_n = f(x_n, y_n)$.

If we set $A(t) = a_0 + a_1t + ... + a_kt^k$ and $B(t) = b_0 + b_1t + ... + b_kt^k$, then we have the following linear multistep method

$$a_0y_n + a_1(E-I)y_n + ... + a_k(E-I)^ky_n = h [b_0f_n + b_1(E-I)f_n + ... + b_k(E-I)^kf_n] .$$

Since $(E-I)^i = \sum_{j=o}^{i} C_i^j (-1)^j E^{i-j}$ this can be written in the more familiar form

$$\alpha_0y_n + \alpha_1Ey_n + ... + \alpha_k E^ky_n = h [\beta_0f_n + \beta_1Ef_n + ... + \beta_kE^kf_n]$$

or

$$\alpha_0y_n + \alpha_1y_{n+1} + ... + \alpha_k y_{n+k} = h [\beta_0f_n + \beta_1f_{n+1} + ... + \beta_kf_{n+k}]$$

with

$$\alpha_i = \frac{1}{i!} \sum_{j=o}^{k-i}(-1)^j(j+1)...(j+i)a_{j+i} \quad , \quad \alpha(t) = \alpha_0 + \alpha_1t + ... + \alpha_kt^k$$

$$\beta_i = \frac{1}{i!} \sum_{j=o}^{k-i}(-1)^j(j+1)...(j+i)b_{j+i} \quad , \quad \beta(t) = \beta_0 + \beta_1t + ... + \beta_kt^k$$

and the convention that $(j+1) ... (j+i) = 1$ if $i = 0$.

Let L be the operator

$$L = \sum_{i=0}^{k} \alpha_i \, E^i - h \sum_{i=0}^{k} \beta_i \, E^i D.$$

It is well known that the linear multistep method has order p if

$$L \, t^j = 0 \qquad j = 0,...,p$$

which corresponds to [136]

$$R(t) = Log(1+t) + O(t^{p+1}) \qquad\qquad (t \to 0).$$

Clearly $p \le 2k$.

We have
$$L \, t^0 = \alpha(1)$$
$$L \, t = t\alpha(1) + h(\alpha'(1) - \beta(1))$$

and thus the method has order one at least if and only if

$$\alpha(1) = 0$$
$$\alpha'(1) = \beta(1)$$

which are the usual conditions for the consistency.
It is well known that such a method is stable if and only if the zeros of α are in the closed unit disc and the zeros of modulus one are simple.
If $\beta_k = 0$ the multistep method is explicit otherwise it is implicit.

We consider the differential equation $y' = \lambda y$ where $\lambda \in \mathbb{C}$, $Re(\lambda) < 0$ and with the initial condition $y(0) = 1$. The multistep method is said to be A-stable if and only if $\lim_{n \to \infty} y_n = 0$, $\forall \lambda$ with $Re(\lambda) < 0$ and $\forall h > 0$. Let W be the exterior of the closed unit disc in the complex plane $W = \{z \mid |z| > 1\}$, then a linear multistep method is A-stable if and only if $\forall z \in W$, Re $R(z-1) \ge 0$. It is also well known that an explicit linear multistep method cannot be A-stable. Thus we have to look only for approximations R of Log $(1+t)$ whose degree of the numerator is not strictly greater than the degree of the denominator. In that case we have an implicit method. Not all the implicit methods are A-stable since the order p of an A-stable linear multistep method cannot exceed 2. The best A-stable linear multistep method of order 2 (that is with the smallest asymptotic error constant) is the trapezoidal rule

$$y_{n+1} = y_n + \frac{h}{2} (f_n + f_{n+1}).$$

It corresponds to $\alpha(t) = t-1$ and $\beta(t) = (t+1)/2$. Thus $R(t) = 2t/(2+t)$ which is the [1/1] Padé approximant to $\text{Log}(1+t)$ and clearly satisfies the stability condition since $t=1$ is the only zero fo α.

Of course $y(x) = e^{\lambda x}$ is the solution of the Cauchy problem considered in the definition of A-stability and we have

$$L\, e^{\lambda x} = \sum_{i=0}^{k} \alpha_i e^{\lambda(x+ih)} - h\lambda \sum_{i=0}^{k} \beta_i e^{\lambda(x+ih)} = c_{p+1}(h\lambda)^{p+1} e^{\lambda x}(1+0(h\lambda))$$

that is

$$\sum_{i=0}^{k} \alpha_i\, e^{it} - t \sum_{i=0}^{k} \beta_i e^{it} = O(t^{p+1})\ .$$

Since $y(x_{n+1}) = e^{h\lambda} y(x_n)$ we shall write that

$$y_{n+1} = r(h\lambda)\, y_n.$$

Thus $r(t)$ is an approximation of e^t which satisfies

$$\sum_{i=0}^{k} (\alpha_i - t\beta_i)\, r^i(t) = 0.$$

This polynomial has k zeros $r_1(t),\ldots,r_k(t)$ and, moreover, we have

$$[e^t - r_1(t)] \ldots [e^t - r_k(t)] = O(t^{p+1}).$$

But when $t = 0$ we shall have

$$\sum_{i=0}^{k} \alpha_i\, r^i(0) = 0.$$

Since all the zeros of α must be inside the unit disc and those on the unit circle must be simple there is one and only one r_i such that $r(0) = 1$ and thus there is one and only one zero of

$$\alpha(r(t)) - t\beta(r(t)) = 0$$

which satisfies

$$r(t) = e^t + O(t^{p+1}).$$

(see, for example, [86]).
In order for the method to be A-stable this r must be analytic in the left half complex plane and satisfy $|r(it)| \leq 1$, $\forall t \in \mathbf{R}$, $\lim_{|t| \to \infty} |r(t)| \leq 1$.

Of course Padé-type, Padé and partial Padé approximants [30] are candidate for such an r. Some were studied in [16] and [100]. A determinantal formula, similar to those used in the previous sections, for partial Padé approximants is given in [157].

The case of the second order differential equation $y''(x) = f(x, y(x))$ can be treated in a similar way. Since it can be written as

$$D^2 y(x) = f(x, y(x))$$

R must now be a rational approximation of $[\text{Log}(1+t)]^2$ and the differential equation is replaced by the difference equation

$$A(\Delta)y_n = h^2 B(\Delta)f_n.$$

We have

$$\text{Log}^2(1+t) = t^2 - t^3 + \frac{11}{12} t^4 - \frac{7}{12} t^5 + O(t^6).$$

It is easy to see that Numerov's method given by

$$y_{n+2} - 2y_{n+1} + y_n = \frac{h^2}{12} (f_{n+2} + 10 f_{n+1} + f_n)$$

corresponds to

$$R(t) = 12t^2/(12 + 12t + t^2) = \text{Log}^2 (1+t) + O(t^5)$$

which shows that R is the [2/2] Padé approximant of $\text{Log}^2 (1+t)$, and that Numerov's method has order 4.

The study of the stability (called P-stability in this case) and of the so-called phase lag can be conducted via the model equation $y'' = -w^2 y$. Since the solution of this differential equation satisfies

$$y(x_{n+2}) - 2 \cos wh \, y(x_{n+1}) + y(x_n) = 0$$

we shall write that

$$y_{n+2} - 2r(w^2h^2) y_{n+1} + y_n = 0$$

which shows that $r(w^2h^2)$ must be an approximation of cos wh.

The multistep method will be said to be P-stable if $|r(t^2)| < 1$ for real $t^2 > 0$. For a complete study see [47] and the references given herein. More generally if we want to solve an operator equation of form

$$Ay = f$$

and if A has a formal series expansion with respect to an operator B

$$A = a_0I + a_1B + a_2B^2 + ...$$

we can replace A by an approximation $N(B)/D(B)$ and solve the approximate equation

$$N(B)y = D(B)f.$$

5.3 - Approximation of series.

Let c be the linear functional on the space of polynomials defined by $c(x^i) = c_i$ for $i = 0,1,...$ ($c_i = 0$ if $i < 0$). Then we formally have

$$f(t) = c((1-xt)^{-1}) = c_0 + c_1t + c_2t^2 + ...$$

If P is the Hermite interpolation polynomial of $(1-xt)^{-1}$ at the zeros of a given polynomial v_k of degree k, then $c(P(x))$ is a rational function with a numerator of degree k-1 and a denominator of degree k, called a Padé-type approximant of f, denoted by $(k-1/k)_f(t)$ and such that

$$(k-1/k)_f(t) = f(t) + O(t^k).$$

If v_k is the formal orthogonal polynomial of degree k with respect to c, that is $v_k \equiv P_k$ such that $c(x^i P_k(x)) = 0$ for $i = 0,...,k-1$ then $(k-1/k)_f(t)$ is the usual Padé approximant [k-1/k] of f and we have

$$[k-1/k]_f(t) = f(t) + O(t^{2k}).$$

If the zeros of v_k are assumed to be distinct then, from the determinantal formula for P, we have

$$(k-1/k)_f(t) = - \frac{\begin{vmatrix} 0 & c_0 & \dots & c_{k-1} \\ (1-x_1t)^{-1} & 1 & \dots & x_1^{k-1} \\ \dots & \dots & \dots & \dots \\ (1-x_kt)^{-1} & 1 & \dots & x_k^{k-1} \end{vmatrix}}{\begin{vmatrix} 1 & x_1 & \dots & x_1^{k-1} \\ \dots & \dots & \dots & \dots \\ 1 & x_k & \dots & x_k^{k-1} \end{vmatrix}}.$$

Of course if some zeros of v_k coincide $(1-x_it)^{-1}, 1, x_i,\dots,x_i^{k-1}$ have to be replaced by their derivatives up to the multiplicity of the zero minus one.

Let us now generalize by replacing the linear functionals previously used (that is the evaluation functionals of a function and its derivatives at the points x_i) by any linear functionals, that is using the interpolation polynomial P such that $L_i(P) = L_i((1-xt)^{-1})$ for $i = 0,\dots,k-1$. We thus obtain a generalization of Padé-type approximants for which the same notation will be kept although it will not, in general, designate a rational function. We have

$$(k-1/k)_f(t) = - \frac{\begin{vmatrix} 0 & c_0 & \dots & c_{k-1} \\ L_0((1-xt)^{-1}) & L_0(1) & \dots & L_0(x^{k-1}) \\ \dots & \dots & \dots & \dots \\ L_{k-1}((1-xt)^{-1}) & L_{k-1}(1) & \dots & L_{k-1}(x^{k-1}) \end{vmatrix}}{\begin{vmatrix} L_0(1) & \dots & L_0(x^{k-1}) \\ \dots & \dots & \dots \\ L_{k-1}(1) & \dots & L_{k-1}(x^{k-1}) \end{vmatrix}}.$$

$L_i((1-xt)^{-1})$ is a function of t that will be denoted by $f_i(t)$ and we formally have

$$f_i(t) = f_i(0) + f_i(1)t + f_i(2)t^2 + \dots \text{ with } L_i(x^j) = f_i(j).$$

Thus multiplying the second column in the numerator of (k-1/k) by 1, the third by t,..., the last one by t^{k-1} and adding to the first column, we obtain

$(k-1/k)_f(t) = c_0+c_1t+... + c_{k-1}t^{k-1}$

$$- t^k \begin{vmatrix} 0 & c_0 & ... & c_{k-1} \\ L_0(x^k(1-xt)^{-1}) & L_0(1) & ... & L_0(x^{k-1}) \\ ... & ... & ... & ... \\ L_{k-1}(x^k(1-xt)^{-1}) & L_{k-1}(1) & ... & L_{k-1}(x^{k-1}) \end{vmatrix} / D_k^{(o,o)}$$

that is

$$(k-1/k)_f = f(t) + O(t^k)$$

which shows that these new approximants satisfy the same approximation property as the Padé-type approximants, the only (but major) difference being that (k-1/k) is a linear combination of the functions $f_0,...,f_{k-1}$ that is

$$(k-1/k)_f (t) = a_0f_0(t) + ... + a_{k-1}f_{k-1}(t)$$

where the a_i's satisfy

$$a_0L_0(1) + ... + a_{k-1} L_{k-1}(1) = c_0$$

$$\text{-------------------------------------}$$

$$a_0L_0(x^{k-1}) + ... + a_{k-1} L_{k-1}(x^{k-1}) = c_{k-1}.$$

Moreover

$f(t)-(k-1/k)_f(t)=$

$$t^k \begin{vmatrix} c(x^k(1-xt)^{-1}) & c_0 & ... & c_{k-1} \\ L_0(x^k(1-xt)^{-1}) & L_0(1) & ... & L_0(x^{k-1}) \\ ... & ... & ... & ... \\ L_{k-1}(x^k(1-xt)^{-1}) & L_{k-1}(1) & ... & L_{k-1}(x^{k-1}) \end{vmatrix} / D_k^{(o,o)}.$$

Let P_k be the monic biorthogonal polynomial of degree k satisfying

$$L_i(P_k(x)) = 0 \qquad\qquad i = 0,...,k-1.$$

We have

$f(t)-(k-1/k)_f(t)=$

$$t^k c(P_k(x)) + t^{k+1} \begin{vmatrix} c(x^{k+i}(1-xt)^{-i}) & c_0 & \dots & c_{k-1} \\ L_0(x^{k+1}(1-xt)^{-1}) & L_0(1) & \dots & L_0(x^{k-1}) \\ \dots & \dots & \dots & \dots \\ L_{k-1}(x^{k+1}(1-xt)^{-1}) & L_{k-1}(1) & \dots & L_{k-1}(x^{k-1}) \end{vmatrix} / D_k^{(0,0)} .$$

This shows that, in general, the degree of approximation cannot be increased unless

$$a_0 L_0(x^{k+i}) + \dots + a_{k-1} L_{k-1}(x^{k+i}) = c_{k+i}$$

for $i = 0,...,m-1$. If these conditions hold then

$$f(t) - (k-1/k)_f(t) = O(t^{k+m}).$$

In fact, for increasing the order of approximation one has to choose the functionals L_i in a proper way (that is such that this system is satisfied with m=k) and this is exactly what is done in the Padé case when selecting the interpolation points $x_1,...,x_k$ as the zeros of the orthogonal polynomial P_k.

Let us set

$$e_k = (-1)^k \begin{vmatrix} c(1) & L_0(1) & \dots & L_{k-1}(1) \\ \dots & \dots & \dots & \dots \\ c(x^{k-1}) & L_0(x^{k-1}) & \dots & L_{k-1}(x^{k-1}) \\ c & L_0 & \dots & L_{k-1} \end{vmatrix} / D_k^{(0,0)} .$$

Thus, if we set $L_{-1} = c$, the functional e_k is, apart from a multiplying factor, the functional $L_k^{(-1,0)}$ of the previous sections, and we have

$$f(t) - (k-1/k)_f(t) = t^k e_k(x^k(1-xt)^{-1})$$

which is an expression for the error very similar to that for the ordinary Padé-type approximants [17, theorem 1.4, p. 20].
Thus if $e_k(x^{k+i}) = 0$ for $i = 0,...,k-1$ we shall have

$$f(t) - (k-1/k)_f(t) = O(t^{2k})$$

which is a generalization of Padé approximants and, in that case, $(k-1/k)$ will be denoted $[k-1/k]$. We have

$$(-1)^k e_k = c - (b_0 L_0 + \dots + b_{k-1} L_{k-1})$$

and the conditions for increasing the order of approximation are

$$b_0 L_0(x^{k+i}) + ... + b_{k-1} L_{k-1}(x^{k+i}) = c_{k+i} \qquad i = 0,...,k-1$$

which is exactly the previous system and shows that $b_j = a_j$ for $j = 0,...,$ $k-1$. Thus $L_0,...,L_{k-1}$ have to be chosen in order to satisfy this system if we want an approximation of order $2k$.

Let us now assume that, instead of being a formal power series, f is a series of functions

$$f(t) = c_0 g_0(t) + c_1 g_1(t) + c_2 g_2(t) + ...$$

Let $G(x,t)$ be the generating function of the g_i's defined by

$$G(x,t) = g_0(t) + x\, g_1(t) + x^2\, g_2(t) + ...$$

and let us replace, in our definition of Padé-type approximants, the function $(1-xt)^{-1}$ (which is the generating function of $g_i(t) = t^i$) by $G(x,t)$. We obtain exactly the same type of results

$$(k-1/k)_f(t) = - \frac{\begin{vmatrix} 0 & c_0 & ... & c_{k-1} \\ L_0(G(x,t)) & L_0(1) & ... & L_0(x^{k-1}) \\ ... & ... & ... & ... \\ L_{k-1}(G(x,t)) & L_{k-1}(1) & ... & L_{k-1}(x^{k-1}) \end{vmatrix}}{\begin{vmatrix} L_0(1) & ... & L_0(x^{k-1}) \\ ... & ... & ... \\ L_{k-1}(1) & ... & L_{k-1}(x^{k-1}) \end{vmatrix}} .$$

Setting $f_i(t) = L_i(G(x,t))$ we have

$$(k-1/k)_f(t) = c_0 g_0(t) + ... + c_{k-1} g_{k-1}(t)$$

$$- \begin{vmatrix} 0 & c_0 & ... & c_{k-1} \\ L_0(x^k G_k(x,t)) & L_0(1) & ... & L_0(x^{k-1}) \\ ... & ... & ... & ... \\ L_{k-1}(x^k G_k(x,t)) & L_{k-1}(1) & ... & L_{k-1}(x^{k-1}) \end{vmatrix} / D_k^{(0,0)}$$

$$= f(t) + O(g_k(t))$$

with $G_k(x,t) = g_k(t) + x\ g_{k+1}(t) + x^2\ g_{k+2}(t) + \ldots$ and where $O(g_k(t))$ designates a series beginning with the term $g_k(t)$.

We thus obtain a generalization of Padé-type approximants for series of functions as defined in [17] and studied in [137]. $(k-1/k)$ is again a linear combination of $f_0,\ldots,f_{k-1}$ as above where the a_i's satisfy the same system of equations. Since this system does not depend on G the a_i's are the same as for $(k-1/k)_f$ when f is a power series with the same coefficients c_i. Thus for a series of functions we only have to replace the f_i's by the new ones, a property already used by Van Rossum [163]. Thus there are two stages for obtaining an approximation of

$$f(t) = c(G(x,t)).$$

First $G(x,t)$ is replaced by its interpolation polynomial P such that

$$L_i(P) = L_i(G(x,t)) \qquad \text{for } i = 0,\ldots,k-1.$$

$(k-1/k)_f(t) = c(P(x))$ is an approximation of f such that

$$(k-1/k)_f(t) = f(t) + O(g_k(t)).$$

Moreover $(k-1/k)$ has the form

$$(k-1/k)_f(t) = a_0 L_0(G(x,t)) + \ldots + a_{k-1} L_{k-1}(G(x,t))$$

where the a_i's satisfy

$$a_0 L_0(x^i) + \ldots + a_{k-1} L_{k-1}(x^i) = c(x^i) \qquad \text{for } i = 0,\ldots,k-1,$$

that is $L_k^*(x^i) = 0$ for $i = 0,\ldots,k-1$ with $L_k^* = a_0 L_0 + \ldots + a_{k-1} L_{k-1} - c$.
Now if we want to increase the order of approximation we shall choose $L_0,\ldots,L_{k-1}$ such that

$$a_0 L_0(x^i) + \ldots + a_{k-1} L_{k-1}(x^i) = c(x^i) \qquad \text{for } i = k,\ldots,2k-1.$$

In that case $c(P(x))$ will be denoted by $[k-1/k]$ and we have

$$[k-1/k]_f(t) = f(t) + O(g_{2k}(t)).$$

In the first case if G is a polynomial of degree at most $k-1$ in x then $(k-1/k)$ is identically f. In the second case if G is a polynomial of degree at most $2k-1$ in x then $[k-1/k]$ is identically f. This is exactly the well known property of interpolatory quadrature formulae when, in the first case, the functionals L_i are arbitrary and, in the second case, they are chosen in an optimal way thus leading to Gaussian quadrature methods.

In the second case let now P_k be the monic orthogonal polynomial of degree k with respect to c. We set

$$P_k(x) = b_0 + b_1 x + ... + b_{k-1} x^{k-1} + x^k.$$

Since $a_0 L_0(x^i) + ... + a_{k-1} L_{k-1}(x^i) = c(x^i)$ for $i = 0,...,2k-1$ we have, multiplying equation i by b_0, equation i+1 by b_1,..., equation i+k-1 by b_{k-1}, equation i+k by 1 and adding

$$c(x^i P_k(x)) = a_0 L_0(x^i P_k(x)) + ... + a_{k-1} L_{k-1}(x^i P_k(x)) = 0 \quad i = 0,...,k-1.$$

In the usual Padé case this is obviously true since L_j is the evaluation functional at the point x_j which is a zero of P_k (or the evaluation functional of one of its derivative is x_j is not simple). In the generalized case studied here, the difficulty is to find such L_i's.

For arbitrary p and q the Padé-type approximants (p/q) and the Padé approximants [p/q] can be constructed from (k - 1/k) and [k-1/k] since, for $n \geq 0$

$$(n+k/k)_f = c_0 + ... + c_n t^n + t^{n+1}(k-1/k)_{f_n} \text{ with } f_n(t) = c_{n+1} + c_{n+2} t +$$

$$(k/n+k)_f = t^{-n+1}(n+k-1/n+k)_{\bar{f}_n} \text{ with } \bar{f}_n(t) = 0 + ... + 0 t^{n-2} + c_0 t^{n-1} + c_1 t^n + ...$$

Let us generalize one step further. Instead of taking $1, x,...,x^{k-1}$ as a basis of P_{k-1}, let us now take $u_0(x), u_1(x),...,u_{k-1}(x)$ where u_i is a polynomial of degree i. Using the notations of the introduction, let $R_{k-1} \in Span (u_0,...,u_{k-1})$ satisfy the interpolation conditions

$$L_i(R_{k-1}) = L_i((1-xt)^{-1}) \quad i = 0,...,k-1.$$

$c(R_{k-1}(x)) = (k-1/k)_f(t)$ will be an approximation of $c((1-xt)^{-1}) = f(t)$ and we shall have

$$(k-1/k)_f(t) = - \frac{\begin{vmatrix} 0 & c(u_0(x)) & ... & c(u_{k-1}(x)) \\ L_0((1-xt)^{-1}) & L_0(u_0(x)) & ... & L_0(u_{k-1}(x)) \\ ... & ... & ... & ... \\ L_{k-1}((1-xt)^{-1}) & L_{k-1}(u_0(x)) & ... & L_{k-1}(u_{k-1}(x)) \end{vmatrix}}{\begin{vmatrix} L_0(u_0(x)) & ... & L_0(u_{k-1}(x)) \\ ... & ... & ... \\ L_{k-1}(u_0(x)) & ... & L_{k-1}(u_{k-1}(x)) \end{vmatrix}}.$$

But, as we saw in section 3

$$R_{k-1} = \sum_{i=0}^{k-1} L_i^* ((1-xt)^{-1}) \, u_i(x)$$

and we also have

$$(k-1/k)_f(t) = \sum_{i=0}^{k-1} c(u_i(x)) \, L_i^* ((1-xt)^{-1}).$$

This is exactly the approach followed by Prévost [155] when he expanded $(1-xt)^{-1}$ into a series of polynomials

$$(1-xt)^{-1} = \sum_{i=0}^{\infty} f_i^* (t) \, u_i(x)$$

and truncated it after u_{k-1}. In that case L_i^* is the functional which associates to a function g the coefficient of u_i in its expansion. Prévost treated the cases where u_i is the Chebyshev polynomial of first or second kind.

If f is a series of functions, $(1-xt)^{-1}$ has to be replaced by $G(x,t)$ and then expanded into a series of polynomials u_i.

5.4 - Biorthogonal polynomials.

The words "biorthogonal polynomials" have been used for a long time and they cover different objects having some connections and which can be put into the general framework described above.

This concept seems to have been studied for the first time by Didon in 1869 [62]. He considered two sets of polynomials $\{U_{k,r}\}$ and $\{V_{k,r}\}$ of degree k with respect to x^r such that if $n \neq k$

$$\int_0^1 U_{k,r} \, V_{n,r} \, dx = 0.$$

Extensions of this notion, obtained by introducing a positive weight function in the integral and changing the interval of integration, were studied by Deruyts in 1886 [61].

This concept of biorthogonal polynomials was further generalized by several authors (which are not listed here) and, among them, by Konhauser [117] who considered the case

$$\int_a^b P_m(x) \, Q_n(x) \, w(x) \, dx = 0 \qquad m \neq n$$

where P_m and Q_n are polynomials of degree m and n in r and s respectively, r and s being polynomials of given degrees.

Formal biorthogonal polynomials were considered by Van Rossum [164] where reference to previous works can be found. More recently, they received a combinatorial interpretation [116] generalizing that of Viennot for the usual orthogonal polynomials [185]. Biorthogonal Laurent polynomials are studied in [96].

Another type of biorthogonal polynomials was proposed in [107]. We shall now study it in more details. We set

$$I_k(\mu) = \int_a^b x^k \, d\alpha(x, \mu).$$

A family of polynomials $\{P_k\}$ is said to be biorthogonal if $\forall k$, P_k has the exact degree k and satisfies

$$\int_a^b P_k(x) \, d\alpha(x, \mu_i) = 0 \qquad \text{for } i = 0,...,k-1.$$

Adjacent families of biorthogonal polynomials $\{P_k^{(n)}\}$ can be defined similarly by

$$\int_a^b x^n \, P_k^{(n)}(x) \, d\alpha(x,\mu_i) = 0 \qquad \text{for } i = 0,...,k-1.$$

Such biorthogonal polynomials have applications in designing multistep methods for integrating ordinary differential equations [108], in rational approximation of Stieltjes functions [112], in the study of the zeros of transformed polynomials [109] and in numerical quadrature [106]. Their theory has been studied in [110]. The same type of biorthogonal polynomials, but in the formal case, was also considered in [26] under the name of multi-orthogonal polynomials which seems to be more appropriate since we only consider one family of polynomials satisfying

$$L_i(P_k) = 0 \qquad \text{for } i = 0,...,k-1$$

where the L_i's are linearly independent functionals. The case of biorthogonal polynomials corresponds to

$$L_i(x^k) = I_k(\mu_i).$$

In [111], adjacent families of biorthogonal polynomials were proved to satisfy a recurrence relationship. This relation is a direct application of the E-algorithm with $S_n = x^n$ and $g_i(n) = I_n(\mu_i)$ as shown in [33] (it can also be obtained from the H-algorithm). Thanks to the theory of section 4 and the recurrence relations of sections 4.1 and 4.2 we can generalize biorthogonal (or multi-orthogonal) polynomials one step further and give new recurrence relationships.

Let us assume that E is a commutative algebra and let us set $x_i = x^i$ with $x \in E$. Then

$$x_n^{(i,j)} = x^j \begin{vmatrix} L_i(x_j) & ... & L_i(x_{j+n}) \\ ... & ... & ... \\ L_{i+n-1}(x_j) & ... & L_{i+n-1}(x_{j+n}) \\ 1 & ... & x^n \end{vmatrix} \Bigg/ \begin{vmatrix} L_i(x_j) & ... & L_i(x_{j+n-1}) \\ ... & ... & ... \\ L_{i+n-1}(x_j) & ... & L_{i+n-1}(x_{j+n-1}) \end{vmatrix}.$$

Let us define the polynomial $P_n^{(i,j)}$ by

$$x_n^{(i,j)} = x^j P_n^{(i,j)}(x).$$

Then we have the biorthogonality property

$$L_p(x^j P_n^{(i,j)}(x)) = 0 \qquad \text{for } p = i,...,i+n-1$$

which reduces to the biorthogonality of Iserles and Nørsett when $i = 0$.

F_5 immediately gives

$$P_n^{(i,j)}(x) = x \ P_{n-1}^{(i,j+1)}(x) - \frac{L_{i+n-1}(x^{j+1} P_{n-1}^{(i,j+1)}(x))}{L_{i+n-1}(x^j P_{n-1}^{(i,j)}(x))} \ P_{n-1}^{(i,j)}(x)$$

which is exactly the recurrence relation obtained in [111].

The other formulae of section 4.1 provide new recurrence relations for the $P_n^{(i,j)}$'s. Thus we obtain from F_4, F_6 and F_7 respectively

$$P_n^{(i,j)}(x) = x \, P_{n-1}^{(i+1,j+1)}(x) - \frac{L_i(x^{j+1} \, P_{n-1}^{(i+1,j+1)}(x))}{L_i(x^j \, P_{n-1}^{(i+1,j)}(x))} \, P_{n-1}^{(i+1,j)}(x)$$

$$P_n^{(i,j)}(x) = P_n^{(i+1,j)}(x) - \frac{L_i(x^j \, P_n^{(i+1,j)}(x))}{L_i(x^j \, P_{n-1}^{(i+1,j)}(x))} \, P_{n-1}^{(i+1,j)}(x)$$

$$P_n^{(i+1,j)}(x) = P_n^{(i,j)}(x) - \frac{L_{i+n}(x^j \, P_n^{(i,j)}(x))}{L_{i+n}(x^j \, P_{n-1}^{(i+1,j)}(x))} \, P_{n-1}^{(i+1,j)}(x) \, .$$

When the upper indexes i and j are fixed, the bordering method can be used to compute recursively the sequence $P_0^{(i,j)}$, $P_1^{(i,j)}$, $P_2^{(i,j)}$,... as described in [32]. The multistep formulae given in section 4.2 can also be applied to adjacent families of biorthogonal polynomials.

Let us now assume that the linear functionals satisfy

$$L_i(x^{r+m}) = L_{i+md}(x^r) \qquad i = 0,...,d-1.$$

If $n = r+md$ with $0 \le r < d$, the previous biorthogonality relations

$$L_p(x^j \, P_n^{(o,j)}(x)) = 0 \qquad \text{for } p = 0,...,n-1$$

become

$$L_p(x^{k+j} \, P_n^{(o,j)}(x)) = 0 \quad \text{for } k = 0,...,m-1 \text{ and } p=0,...,d-1$$

$$L_p(x^{m+j} \, P_n^{(o,j)}(x)) = 0 \qquad \qquad \text{for } p = 0,...,r-1.$$

Thus the biorthogonal polynomials $P_n^{(o,j)}$ are identical to the vector orthogonal polynomials of Van Iseghem [103]. Such polynomials satisfy a recurrence relation with $d+2$ terms (when $d = 1$, the usual three-terms recurrence relationship for orthogonal polynomials is recovered). Moreover [105]

$$L_0(x^{k+j} \; P_n^{(o,j)}(x)) = 0 \quad \text{for } n \geq kd+1 \text{ and } k \geq 0$$

which shows that these polynomials are $1/d$-orthogonal with respect to the functional L_0, a notion introduced by Maroni [131] (a formalism for their study is given in [132]).

Relations between adjacent families of vector orthogonal polynomials are given in [103], one of them reducing to the relation due to Iserles and Nørsett since vector orthogonal polynomials are a particular case of the biorthogonal ones. Vector orthogonal polynomials can also be computed by a generalization of the qd-algorithm. They satisfy an extension of Favard's theorem and their zeros have been studied [105]. Such polynomials have applications in vector Padé approximants which are rational approximants with a common denominator which approximate simultaneously d formal power series [101, 102, 105].

As pointed out in [112], the denominators of the simultaneous approximants of de Bruin [41] are also related to biorthogonal polynomials. See [33] for more details and [42] for a generalization.

Vector orthogonal polynomials can be generalized by taking the first upper index i different from zero.

Orthogonal Laurent polynomials were introduced by Jones and Thron [115] in connection with two point Padé approximants and T continued fractions. A Laurent polynomial is an expression of the form

$$P(x) = \sum_{j=k}^{m} a_j x^j \quad \text{with } -\infty < k \leq m < +\infty.$$

We shall denote their set by R and we set

$$R_{2m} = \text{Span}(x^{-m}, x^{-m+1},...,x^{-1}, 1, x,...,x^m)$$

$$R_{2m-1} = \text{Span}(x^{-m},..., x^{m-1}).$$

Let c be the linear functional on R defined by its moments

$$c_i = c(x^i) \qquad\qquad i = 0, \pm 1, \pm 2...$$

We consider the monic Laurent polynomials R_{2n} and R_{2n+1} (n = 0,1,...) defined by

$$R_{2n}(x) = \begin{vmatrix} c_{-2n} & \cdots & c_0 \\ \cdots & \cdots & \cdots \\ c_{-1} & \cdots & c_{2n-1} \\ x^{-n} & \cdots & x^n \end{vmatrix} \Big/ H_{2n}^{(-2n)}$$

$$R_{2n+1}(x) = \begin{vmatrix} c_{-2n-1} & \cdots & c_0 \\ \cdots & \cdots & \cdots \\ c_{-1} & \cdots & c_{2n} \\ x^{-n-1} & \cdots & x^n \end{vmatrix} \Big/ H_{2n+1}^{(-2n-1)} .$$

They satisfy

$$c(x^i R_{2n}(x)) = 0 \qquad\qquad i = -n,\dots,n-1$$

$$c(x^i R_{2n+1}(x)) = 0 \qquad\qquad i = -n,\dots,n.$$

The family $\{R_n\}$ is called a family of orthogonal Laurent polynomials with respect to the functional c. They are known to satisfy the recurrence relations

$$R_{2n}(x) = (A_{2n}x + B_{2n})R_{2n-1}(x) - C_{2n}R_{2n-2}(x)$$

$$R_{2n+1}(x) = (A_{2n+1}x^{-1} + B_{2n+1})R_{2n}(x) - C_{2n+1}R_{2n-1}(x)$$

with $R_0(x) = 1$ and $R_1(x) = 1 - c_0 x^{-1}$.

Multiplying orthogonal Laurent polynomials by the suitable power of x leads to ordinary polynomials; thus let us set

$$V_{2n}(x) = x^n R_{2n}(x)$$

$$V_{2n+1}(x) = x^{n+1} R_{2n+1}(x).$$

We have

$$c(x^i V_{2n}(x)) = 0 \qquad\qquad i = -2n,\dots,-1$$

$$c(x^i V_{2n+1}(x)) = 0 \qquad\qquad i = -2n-1,\dots,-1$$

which can be written as

$$c^{(-2n)}(x^i V_{2n}(x)) = 0 \qquad\qquad i = 0,\dots,2n-1$$

$$c^{(-2n-1)}(x^i V_{2n+1}(x)) = 0 \qquad i = 0,\dots,2n.$$

Thus V_{2n} and V_{2n+1} are members of two adjacent families of orthogonal polynomials since V_{2n} is identical to $P_{2n}^{(-2n)}$ and V_{2n+1} to $P_{2n+1}^{(-2n-1)}$ and therefore their theory fits into our framework as remarked by Draux [64].

Padé approximants for Laurent series were introduced by Gragg [83], they are called Laurent-Padé approximants. Of course a Laurent series can be splitted into negative and positive powers of the variable and thus there is a strong connection with two-point Padé approximants and orthogonal Laurent polynomials, a connection fully exploited and developed in [43] (see also [81]). Such approximants (and Laurent orthogonal polynomials) have applications ranging from stochastic processes, time series analysis, signal processing, linear systems theory and inverse scattering [43]. There are also connections with polynomials orthogonal on the unit circle.

Let $f_1,...,f_N$ be formal power series

$$f_i(t) = f_0^i + f_1^i t + f_2^i t^2 + ...$$

The Padé-Hermite approximation problem consists in finding the polynomials $P_1,...,P_N$ of respective degrees $n_1,...,n_N$ such that

$$f_1(t)P_1(t) + ... + f_N(t) P_N(t) = O(t^{s+N-1})$$

with $s = n_1 + ... + n_N$. This problem contains the usual Padé approximation problem ($N=1$), quadratic approximation ($N=3$, $f_1 = f$, $f_2 = f^2$, $f_3 = 1$) which was introduced by Shafer [169] (see also [44]) and D-log approximation of Baker [5] ($N=3$, $f_1 = f$, $f_2 = f'$, $f_3 = 1$).

Extending the usual Padé case, it was showed in [5] how to relate Padé-Hermite approximants with an extension of orthogonal polynomials called vector orthogonal polynomials (with no relations to the previous ones). These vector orthogonal polynomials have the representation

$V_i(x) =$

$$\begin{vmatrix} f_0^1 & \cdots & f_{n_1}^1 & \cdots & f_0^i & \cdots & f_{n_i}^i & \cdots & f_0^N & \cdots & f_{n_N}^N \\ \cdots & \cdots & \cdots & \cdots & \cdots & \cdots & \cdots & \cdots & \cdots & \cdots & \cdots \\ f_{s+N-2}^1 & \cdots & f_{s+N-2+n_1}^1 & \cdots & f_{s+N-2}^i & \cdots & f_{s+N-2+n_i}^i & \cdots & f_{s+N-2}^N & \cdots & f_{s+N-2+n_N}^N \\ 0 & \cdots & 0 & \cdots & 1 & \cdots & x^{n_i} & \cdots & 0 & \cdots & 0 \end{vmatrix}$$

Let c^i be the linear functional defined by

$$c^i(x^k) = f_k^i \qquad k = 0,1,\dots\; ; i = 1,\dots,N.$$

Then the polynomials V_i satisfy the orthogonality relation

$$\sum_{i=1}^{N} c^i(x^k V_i(x)) = 0 \qquad k = 0,\dots,s+N+2.$$

Some recursive methods for the computation of these polynomials were given in [65] (see also [60, 153]). Of course V_i depends on $n_1,\dots,n_N$.

Let us now assume that $n_1 = \dots = n_N = n$ and let $P_n(x)$ be the vector with components $V_1(x),\dots,V_N(x)$. Since the V_i are defined apart from an arbitrary non-zero multiplying factor, we have

$$P_n(x) = \begin{vmatrix} f_0^1 & \dots & f_0^N & \dots & f_n^1 & \dots & f_n^N \\ \dots & \dots & \dots & \dots & \dots & \dots & \dots \\ f_{N(n+1)-2}^1 & \dots & f_{N(n+1)-1}^N & \dots & f_{N(n+1)-2+n}^1 & \dots & f_{N(n+1)-2+n}^N \\ I & & & \dots & & & x^n I \end{vmatrix}$$

where I is the $N \times N$ identity matrix.

Let c be the vector of functionals $c^1,\dots,c^N$. If Q is a vector of polynomials with components $Q_1,\dots,Q_N$ we shall make use of the notation

$$c(Q(x)) = \sum_{i=1}^{N} c^i(Q_i(x)).$$

Thus we have

$$c(x^k P_n(x)) = 0 \qquad k = 0,\dots,N(n+1)-2,$$

or

$$L_k(P_n(x)) = 0 \qquad k = 0,\dots,N(n+1)-2$$

if L_k is defined by

$$L_k(Q(x)) = c(x^k Q(x)) = \sum_{i=1}^{N} c^i(x^k Q_i(x)).$$

Going from P_n to P_{n+1} needs the introduction of N new rows and columns in the above determinantal expression, an introduction which can be done step by step. For that, let us define the intermediate polynomials $P_n^{(i)}$ by adding the first i new rows and columns contained in P_{n+1}. Thus $P_n^{(o)}$ is identitcal to P_n and $P_n^{(N)}$ to P_{n+1}. Moreover

$$c(x^k \ P_n^{(i)}(x)) = 0 \qquad\qquad k = 0,...,N(n+1)\text{-}2+i.$$

An interesting open question would be to see if the recurrence relations of sections 4.1 and 4.2 could be used to compute these polynomials.

Several other possible extensions of the notion of orthogonality for polynomials can also be studied in our framework. For example if $\{w_i\}$ is a given family of linearly independent polynomials, one can look for the family $\{P_k\}$ such that

$$c(w_i(x) \ P_k(x)) = 0 \qquad\qquad \text{for } i = 0,...,k\text{-}1.$$

The usual orthogonal polynomials are recovered if $w_i(x) = x^i$. The case $w_o(x) = 1$, $w_i(x) = (x\text{-}x_i) \ w_{i\text{-}1}(x)$ leads to what can be called multipoint orthogonal polynomials.

Another interesting case is that of Stieltjes' polynomials. Let $\{P_k\}$ be the family of formal orthogonal polynomials with respect to c. The polynomial V_{k+1}, of degree k+1, satisfying

$$c(x^i P_k(x) \ V_{k+1}(x)) = 0 \qquad\qquad \text{for } i = 0,...,k$$

is called the Stieltjes' polynomial of degree k+1. If we define the functional L_k by

$$L_k(p(x)) = c(P_k(x) \ p(x))$$

then

$$L_k(x^i V_{k+1}(x)) = 0 \qquad\qquad i = 0,...,k$$

which shows that V_{k+1} is the polynomial of degree k+1 belonging to the family of formal orthogonal polynomials with respect to L_k (which depends on k). Stieltjes' polynomials have important applications in

Gaussian quadrature methods [78, 137] and Padé approximation [27]. They have also been studied in [156] from the formal viewpoint.

Let us now define what can be called orthogonal polynomials in the least squares sense : a monic polynomial P_k of degree k such that

$$d^2 = \sum_{i=0}^{m} [c(x^i P_k(x))]^2$$

is minimum, where $m \geq k-1$. Writing $P_k(x) = a_0 + ... + a_{k-1}x^{k-1} + x^k$ the a_i's are solution of the linear system

$$a_0 \sum_{i=0}^{m} c_i c_{i+j} + ... + a_{k-1} \sum_{i=0}^{m} c_{i+k-1}c_{i+j} + \sum_{i=0}^{m} c_{i+k}c_{i+j} = 0 \qquad j = 0,...,k-1.$$

Setting $\gamma_n = (c_n,...,c_{n+m})^T$, the system writes
$$a_0(\gamma_0, \gamma_j) + ... + a_{k-1}(\gamma_{k-1}, \gamma_j) + (\gamma_k, \gamma_j) = 0 \qquad j = 0,...,k-1.$$

Defining the linear functionals L_i by

$$L_i(x^j) = (\gamma_i, \gamma_j)$$

we have

$$L_i(P_k(x)) = 0 \qquad i = 0,...,k-1$$

which shows that these least squares orthogonal polynomials also fit into the general framework of biorthogonality. Such polynomials could be useful in the definition of Padé approximants in the least squares sense that is rational fractions with a numerator of degree p and a denominator of degree q such that their series expansion $d_0 + d_1t + d_2t^2 + ...$ be such that

$$\sum_{i=0}^{m} (d_i - c_i)^2$$

be minimum $(m \geq p+q)$.
All these notions deserve further study.

5.5 - Statistics and least squares.

There are obviously many connections between statistics and biorthogonality. For example orthogonal expansions and the theory of reproducing kernel Hilbert spaces play a central role in time series analysis that is "the extraction, detection and prediction of signals in the presence of noise" as stated by Parzen [152]. Many papers on this problem were gathered in [187]. Other examples are given by the multivariate normal distribution, the computation of partial correlation coefficients, some special covariance and correlation structures arising in statistical applications, the chi-squared and Wishart distributions, and the Cramér-Rao inequality where Schur complements (that is ratios of determinants similar to ours) have many applications described by Ouellette [150]. Recently biorthogonalization was used for the least-square linear prediction of any statistically dependent random variable and it provides an extension of Slepian's model for Gaussian noise conditioned on any number of derivatives [10]. It is also known that least squares approximation and some estimation problems in statistics have common aspects [58, p. 126] (see also some of the papers contained in [187] and, in particular that of Parzen [151]). On the other hand the problem of optimal linear approximation in a reproducing kernel Hilbert space can be treated by introducing a Gaussian measure and then using well known techniques of probability theory and statistics to obtain properties of the function from the given data [120] (see also [119]).

Although it would be very much useful, our aim in this section is not to rephrase all these results in a common language but it is to give an application of biorthogonality to the computation of the coefficient of correlation and to use this coefficient to chose between several extrapolation procedures for a given sequence.

Let x, y, $x_1,...,x_k$ be random variables. We shall first recall some well known results (see, for example, [63] or [73]). We shall denote by $E(y)$ or by $\bar{y}$ the expectation (mean value) of y and we shall set

$$\text{cov } (xy) = E((x-\bar{x})(y-\bar{y})) = E(xy) - \bar{x}\bar{y}$$

$$\text{var } x = \text{cov}(xx) = E((x-\bar{x})^2) = E(x^2) - \bar{x}^2.$$

We shall define the multiple correlation coefficient of y and $x_1,...,x_k$ by

$$\rho_k = [(a_k, C_k^{-1} a_k)/\text{var } y]^{1/2}$$

with
$$a_k = (\text{cov}(yx_1),...,\text{cov}(yx_k))^T$$

$$C_k = \begin{pmatrix} \text{var } x_1 & \text{cov}(x_1x_2) & ... & \text{cov}(x_1x_k) \\ \text{cov}(x_2x_1) & \text{var } x_2 & ... & \text{cov}(x_2x_k) \\ ... & ... & ... & ... \\ \text{cov}(x_kx_1) & \text{cov}(x_kx_2) & ... & \text{var } x_k \end{pmatrix}.$$

When k=1, ρ_1 is called the linear correlation coefficient.

Thus from Schur's formula

$$(a_k, C_k^{-1} a_k) = - \begin{vmatrix} 0 & \text{cov}(yx_1) & ... & \text{cov}(yx_k) \\ \text{cov}(x_1y) & \text{cov}(x_1x_1) & ... & \text{cov}(x_1x_k) \\ ... & ... & ... & ... \\ \text{cov}(x_ky) & \text{cov}(x_kx_1) & ... & \text{cov}(x_kx_k) \end{vmatrix} / |C_k|$$

which shows that ρ_k can be recursively computed by the algorithms developed in sections 4.1 and 4.2.

As its name indicates ρ_k measures the correlation between y and $x_1,...,x_k$ since we have the following property

Property 1.

Let a, $b_1,...,b_k$ *be constants. If* $y = a+b_1x_1+...+b_kx_k$ *then* $\rho_k = 1$.

Proof :

Let $b = (b_1,...,b_k)^T$. Then

$$C_k b = \begin{pmatrix} \text{cov}(x_1 \sum_{i=1}^{k} b_i x_i) \\ .... \\ \text{cov}(x_k \sum_{i=1}^{k} b_i x_i) \end{pmatrix} = \begin{pmatrix} \text{cov}(x_1(y-a)) \\ .... \\ \text{cov}(x_k(y-a)) \end{pmatrix}.$$

But $\text{cov}(x_i(y-a)) = \text{cov}(x_iy)$ since a is a constant. Thus

$$C_k b = a_k$$

and

$$(a_k, C_k^{-1} a_k) = \sum_{i=1}^{k} b_i \, cov(x_i y) = cov(y \sum_{i=1}^{k} b_i x_i) = cov(y(y-a))$$

$$= cov(yy) = var \ y$$

which shows that $\rho_k = 1$. ◇

Since a is a constant, $var(y+a) = var \ y$ and the b_i's which minimize $var(y - a - b_1 x_1 - ... - b_k x_k)$ are the same that those minimizing $var(y - b_1 x_1 - ... - b_k x_k)$. From the proof of property 1, they are given by $b = C_k^{-1} a_k$. Moreover

$$var \ (y - \sum_{i=1}^{k} b_i \, x_i) = \inf_{d \in R^k} var \ (y - \sum_{i=1}^{k} d_i \, x_i).$$

Since the b_i's have been obtained as the solution of the preceding linear system, we have

$$a = E(y - b_1 x_1 - ... - b_k x_k) = \bar{y} - b_1 \bar{x}_1 - ... - b_k \bar{x}_k$$

or, equivalently (after some manipulations which are omitted)

$$a + b_1 \bar{x}_1 + ... + b_k \bar{x}_k = \bar{y}$$

$$a\bar{x}_1 + b_1 \, E(x_1 x_1) + ... + b_k E(x_1 x_k) = E(x_1 y)$$

$$a\bar{x}_k + b_1 \, E(x_k x_1) + ... + b_k E(x_k x_k) = E(x_k y) \ .$$

Comparing with the system solved in [20], shows the connection with least squares extrapolation by the E-algorithm.

We have the

Property 2.

    The multiple correlation coefficient of y and $b_1 x_1 + ... + b_k x_k$ is maximum for $b = C_k^{-1} a_k$. For this choice it is equal to ρ_k.

The difference between properties 1 and 2 must be clearly understood. In property 1, $b_1,...,b_k$ are arbitrary constants and the multiple correlation coefficient is involved. In property 2, $b_1,...,b_k$ are fixed constants and the linear correlation coefficient is used. Its value is maximal and equal to ρ_k for $b = C_k^{-1} a_k$. In that case the quality of the approximation can be measured by

$$\text{var } \varepsilon_k = \text{var } (y - b_1 x_1 - ... - b_k x_k)$$

and it is easy to prove that

Property 3.

$$\text{var } \varepsilon_k = \text{var } y - (a_k, C_k^{-1} a_k) = \inf_{d \in R^k} \text{var } (y - \sum_{i=1}^{k} d_i x_i).$$

Thus

Property 4.

$$d_k = 1 - \rho_k^2 = \text{var } \varepsilon_k / \text{var } y \geq 0.$$

It follows that $0 \leq \rho_k \leq 1$.

Let us now set

$$d(x,y) = [\text{var}(x-y)]^{1/2} .$$

We have

Property 5.

1°) $d(x,y) \geq 0 \text{ and } d(x,y) = 0 \text{ if } y = a+x \text{ where } a \text{ is a constant.}$
2°) $d(x,y) = d(y,x).$
3°) $d(x,y) \leq d(x,z) + d(z,y).$
4°) $d(x+z, y+z) = d(x,y).$
5°) $d(ax,ay) = |a| d(x,y) \text{ where } a \text{ is a constant.}$
6°) $d(x,y+a) = d(x,y) \text{ where } a \text{ is a constant.}$

This property shows that d is a pseudo-distance. We can obtain a distance by means of the quotient modulo the equivalence relation

$$x \sim x' \Leftrightarrow d(x,x') = 0$$

that is by considering x and x' as identical if and only if $d(x,x') = 0$.

We set

$$N_k = \{x \mid x = \sum_{i=1}^{k} d_i x_i\}$$

and we have

$$d(y, N_k) = \inf_{d \in R^k} d(y, \sum_{i=1}^{k} d_i x_i)$$

$$= \inf_{d \in R^k} [var(y - \sum_{i=1}^{k} d_i x_i)]^{1/2}$$

$$= (var \, \varepsilon_k)^{1/2} = [(1 - \rho_k^2) \, var \, y]^{1/2}$$

$$= d(y, \sum_{i=1}^{k} b_i x_i) \quad \text{with } b = C_k^{-1} a_k \, .$$

Thus $x = \sum_{i=1}^{k} d_i x_i$ is the projection of y on N_k and property 3 is Pythagoras' theorem.

Moreover

$$d^2(y, N_k) = \begin{vmatrix} cov(yy) & cov(yx_1) & ... & cov(yx_k) \\ cov(x_1 y) & cov(x_1 x_1) & ... & cov(x_1 x_k) \\ ... & ... & ... & ... \\ cov(x_k y) & cov(x_k x_1) & ... & cov(x_k x_k) \end{vmatrix} / |C_k|$$

which is a known result if we consider the bilinear form defined by

$$(x|y) = cov(xy).$$

This form is the bilinear form associated with our distance. Indeed we have

$$\|x\| = d(x,0) = (\text{var } x)^{1/2} = (x|x)^{1/2}.$$

Thus

$$(x \mid x) = \text{var } x.$$

A bilinear form is entirely determined by its values on the diagonal. We have

$$2(x|y) = (x+y|x+y) - (x|x) - (y|y)$$

$$= \text{var } (x+y) - \text{var } x - \text{var } y$$

$$= 2 \text{ cov } (xy)$$

and thus

$$(x|y) = \text{cov}(xy).$$

Everything remains valid if E is any linear form such that $E(a) = a$ if a is stationary. We shall assume that $E(x^2) = 0$ if and only if $x = 0$. Then we have

Property 6.   *If $d(x,y) = 0$ then $y = a+x$ where a is a constant.*

Property 7.   *If $d(y, N_k) = 0$ then $y \in N_k$.*

Property 8.   *If $y = a + d_1x_1 + ... + d_kx_k+\varepsilon$ then $d(y, N_k) = d(\varepsilon, N_k)$.*

Proof :

$$d^2(y, N_k) = \text{var } y - (a_k, C_k^{-1} a_k). \text{ Let } d = (d_1,...,d_k)^T.$$

Then

$$C_k d = \begin{pmatrix} \text{cov}(x_1 \sum_{i=1}^{k} d_i x_i) \\ \cdots\cdots\cdots \\ \text{cov}(x_k \sum_{i=1}^{k} d_i x_i) \end{pmatrix} = \begin{pmatrix} \text{cov}(x_1(y-a-\varepsilon)) \\ \cdots\cdots\cdots \\ \text{cov}(x_k(y-a-\varepsilon)) \end{pmatrix} = a_k - a_k'$$

with $a_k' = (\text{cov}(\varepsilon x_1),...,\text{cov}(\varepsilon x_k))^T$. Thus $d = C_k^{-1} a_k - C_k^{-1} a_k'$, that is

$$C_k^{-1} a_k = d + C_k^{-1} a'_k$$

$$(a_k, C_k^{-1} a_k) = (a_k, d) + (a_k, C_k^{-1} a'_k)$$

$$= (a_k, d) + (a'_k + C_k d, C_k^{-1} a'_k)$$

But $C_k = C_k^T$ and then $(C_k d, C_k^{-1} a'_k) = (d, a_k)$.

Thus

$$(a_k, C_k^{-1} a_k) = (a_k, d) + (a'_k, C_k^{-1} a'_k) = (d, a'_k).$$

We have

$$\text{var } y = \text{var } (d_1 x_1 + \dots + d_k x_k + \varepsilon)$$

$$= \text{var } (d_1 x_1 + \dots + d_k x_k) + \text{var } \varepsilon + 2\text{cov}((d_1 x_1 + \dots + d_k x_k)\varepsilon).$$

But

$$\text{cov}((d_1 x_1 + \dots + d_k x_k)\varepsilon) = d_1 \text{cov}(x_1 \varepsilon) + \dots + d_k \text{cov}(x_k \varepsilon)$$

$$= (d, a'_k)$$

and

$$\text{var}(d_1 x_1 + \dots + d_k x_k) = E((d_1 x_1 + \dots + d_k x_k)^2) - (d_1 \bar{x}_1 + \dots + d_k \bar{x}_k)^2$$

$$= \sum_{i,j=1}^{k} d_i d_j (E(x_i x_j) - \bar{x}_i \bar{x}_j)$$

$$= \sum_{i,j=1}^{k} d_i d_j \text{cov}(x_i x_j) = (d, C_k d)$$

$$= (d, a_k) - (d, a'_k).$$

Thus finally

$$d^2(y, N_k) = (d, a_k) - (d, a'_k) + \text{var } \epsilon + 2(d, a'_k) - (a_k, d) - (a'_k, C_k^{-1} a'_k)$$
$$- (d, a'_k)$$

$$= \text{var } \epsilon - (a'_k, C_k^{-1} a'_k) = d^2(\epsilon, N_k). \diamond$$

We previously saw that the bilinear form associated with our distance was $(x|y) = \text{cov}(xy) = E(xy) - \bar{x}\bar{y}$.

Let us now take the bilinear form given by

$$(x,y) = E(xy)$$

and let us set

$$M_k = \{x \mid x = \sum_{i=0}^{k} a_i x_i \text{ with } x_0 = 1\}.$$

We have

$$d^2(y, M_k) = \begin{vmatrix} (y,y) & (y,x_0) & \dots & (y,x_k) \\ (x_0,y) & (x_0,x_0) & \dots & (x_0,x_k) \\ \dots & \dots & \dots & \dots \\ (x_k,y) & (x_k,x_0) & \dots & (x_k,x_k) \end{vmatrix} \Big/ \begin{vmatrix} (x_0,x_0) & \dots & (x_0,x_k) \\ \dots & \dots & \dots \\ (x_k,x_0) & \dots & (x_k,x_k) \end{vmatrix}.$$

But $\forall z$, $(z,x_0) = E(z)$ and thus

$$d^2(y, M_k) = \begin{vmatrix} E(y^2) & E(y) & E(yx_1) & \dots & E(yx_k) \\ E(y) & 1 & E(x_1) & \dots & E(x_k) \\ \dots & \dots & \dots & \dots & \dots \\ E(x_k y) & E(x_k) & E(x_k x_1) & \dots & E(x_k^2) \end{vmatrix} \Big/ \begin{vmatrix} 1 & E(x_1) & \dots & E(x_k) \\ E(x_1) & E(x_1^2) & \dots & E(x_1 x_k) \\ \dots & \dots & \dots & \dots \\ E(x_k) & E(x_k x_1) & \dots & E(x_k^2) \end{vmatrix}.$$

In the numerator let us multiply the second row by $E(y)$ and subtract from the first one. Then we multiply the second row of the numerator by $E(x_1)$ and subtract from the third one, and we do the same for the first and second row in the denominator and so on. We finally obtain the

Property 9. $d^2(y, M_k) = d^2(y, N_k)$.

Thus, from the beginning, it is not necessary to center the variables. Centering the variables just reduces the dimension of the space on which we project since N_k has dimension k and M_k has dimension k+1.

Let x be the projection of y on N_k. We have

$$E(y\text{-}x) = \begin{vmatrix} E(y) & E(x_1) & ... & E(x_k) \\ cov(x_1y) & cov(x^2_1) & ... & cov(x_1x_k) \\ ... & ... & ... & ... \\ cov(x_ky) & cov(x_kx_1) & ... & cov(x^2_k) \end{vmatrix} \Bigg/ \begin{vmatrix} cov(x^2_1) & ... & cov(x_1x_k) \\ ... & ... & ... \\ cov(x_kx_1) & ... & cov(x^2_k) \end{vmatrix}$$

which is also equal to the coefficient of x_0 in the expression giving the projection of y on M_k. Since this coefficient depends on the dimension k, let us denote it by b_k. We have

$$b_k = E(y\text{-}x) = \bar{y} - \sum_{i=1}^{k} (y|x_i)\bar{x}^*_i$$

where x^*_1, x^*_2,.... is obtained by orthogonalizing x_1, x_2,.... with respect to $(.|.)$. We have

$$b_0 = 0$$

$$y_1 = x_1 \qquad\qquad\qquad x^*_1 = y_1/\|y_1\|$$

$$y_k = x_k - \sum_{i=1}^{k-1} (x_k \mid x^*_i)x^*_i \qquad\qquad x^*_k = y_k/\|y_k\|$$

$$b_k = b_{k-1} - (y \mid x^*_k)\bar{x}^*_k \quad , \quad \text{with } \|y_k\|^2 = (y_k \mid y_k) .$$

Moreover

$$d^2(y, M_k) = \|y\|^2 - \sum_{i=1}^{k} |(y, x_i^*)|^2$$

$$y_k = x_k - \text{proj}_{N_{k-1}} x_k$$

and

$$E(y_k) =$$

$$\begin{vmatrix} E(x_k) & E(x_1) & ... & E(x_{k-1}) \\ \text{cov}(x_1 x_k) & \text{cov}(x_1^2) & ... & \text{cov}(x_1 x_{k-1}) \\ ... & ... & ... & ... \\ \text{cov}(x_{k-1} x_k) & \text{cov}(x_{k-1} x_1) & ... & \text{cov}(x_{k-1}^2) \end{vmatrix} \Bigg/ \begin{vmatrix} \text{cov}(x_1^2) & ... & \text{cov}(x_1 x_{k-1}) \\ ... & ... & ... \\ \text{cov}(x_{k-1} x_1) & ... & \text{cov}(x_{k-1}^2) \end{vmatrix}.$$

The relation between this expression and that of $g_{k-1,k}^{(n)}$ in the auxiliary rule of the E-algorithm can be easily seen.
We also have

$$b_k = b_{k-1} - \frac{(y|y_k)}{(y_k|y_k)} \bar{y}_k$$

$$y_k = x_k - \sum_{i=1}^{k-1} \frac{(x_k|y_i)}{(y_i|y_i)} y_i$$

and, since y_k is orthogonal to $y_1,...,y_{k-1}$, then $(y_k|y_k) = (y_k|x_k)$ and it follows that

$$(y_k|y_k) = d^2(x_k, M_{k-1}).$$

Let us now give an application of all these results to sequence extrapolation by the E-algorithm. We already know that this algorithm consists in computing the numbers $E_k^{(n)}$ such that

$$S_{n+i} = E_k^{(n)} + a_1 g_1(n+i) + ... + a_k g_k(n+i) \qquad i = 0,...,k,$$

where the g_i's are given auxiliary sequences (which can depend on (S_n)).
Let

$$N_k = \{(S_n) \mid \forall n, S_n = S + a_1 g_1(n) + ... + a_k g_k(n)\}.$$

It can be proved that $\forall n$, $E_k^{(n)} = S$ if and only if $(S_n) \in N_k$. If we consider the S_n's and the $g_i(n)$'s as realizations of random variables, then the multiple correlation coefficient of (S_n) and the $(g_i(n))$ can be estimated by computing ρ_k with

$$E((S_n)) = \frac{1}{m+1} \sum_{i=o}^{m} S_{n+i}$$

and similarly for the sequence $(g_i(n))$. Of course we must take $m > k$ since, otherwise, ρ_k would be equal to zero. It must be noticed that ρ_k depends on n and m. If $(S_n) \in N_k$, then $\forall n$ and $\forall m > k$, $\rho_k = 1$ that is $d((S_n), N_k) = 0$.

When wanting to accelerate the convergence of a given sequence (S_n) by the E-algorithm, the main practical point is the choice of the auxiliary sequences $(g_i(n))$. In [59], Delahaye introduced a procedure consisting in using simultaneously several extrapolation algorithms (that is several choices for the $(g_i(n))$) and then, at each step n, choosing one result among those given by the various algorithms according to some selection test. Such a new selection test can now be based upon the multiple correlation coefficient :

1 - k and m > k are fixed integers.

2 - We make several choices for the k auxiliary sequences :
$$(g_1^1, ..., g_k^1), (g_1^2, ..., g_k^2), ..., (g_1^p, ..., g_k^p).$$

3 - For a given n we compute the multiple correlation coefficients $\rho_k^1, ..., \rho_k^p$ corresponding to the various sets of auxiliary sequences.

4 - We select the index i such that $\rho_k^i = \max_{1 \le j \le p} \rho_k^j$ and we use the E-algorithm with the auxiliary sequences $g_1^i, ..., g_k^i$. Of course $d((S_n), N_k^i)$ $\min_{1 \le j \le p} d((Sn), N_k^j)$ where

$$N_k^j = \{(S_n) | \forall n, S_n = S + a_1 g_k^j(n) + ... + a_k g_1^j(n)\}$$

5 - Add 1 to n and go to point 3.

In its spirit this selection procedure is very close to another one which will now be described and which is based on a similar technique used in statistics for time series analysis. (I am indebted to Prof. D. Bosq for indicating me this procedure).

In convergence acceleration methods one uses a sample of terms of the sequence to be accelerated, S_n, S_{n+1},...,S_{n+k}, to obtain an approximate value of its limit. Instead of predicting the limit, one can also use the same technique to predict the unknown members of the sequence S_{n+k+1}, S_{n+k+2},... [25]. Let us denote by S'_{n+k+i} the predicted values for $i \geq 1$. If the true values of S_{n+k+1},...,S_{n+m} are known (as was the case in the selection test based upon the multiple correlation coefficient) one can compare them to the predicted values S'_{n+k+1},...,S'_{n+m} by computing

$$\sum_{i=1}^{m-k} (S_{n+k+i} - S'_{n+k+i})^2$$

and choose among the sets of auxiliary sequences $(g_1^i,...,g_k^i)$ the set for which this quantity is minimum.

These two new automatic selection procedures have to be studied from the theoretical and practical points of view. Their possible connection also deserve further research.

There are certainly many other possible connections between statistical methods and extrapolation methods. For example, the ε-algorithm can be used in ARMA models as described in [9]. This algorithm can be considered as a linear filter : we set

$$\varepsilon_n = S - \sum_{i=0}^{k} a_i S_{n-i} \qquad\qquad n \geq k$$

$$I = \sum_{n=k}^{N+k} \varepsilon_n^2$$

and we look for the a_i's minimizing I with $a_0 + ... + a_k = 1$. If $N = k$ then we must have $\varepsilon_n = 0$ for $n = k,...,2k$ that is

$$a_0 S_k + ... + a_k S_0 = S$$
$$\text{-----------------}$$
$$a_0 S_{2k} + ... + a_k S_k = S$$
$$a_0 + ... + a_k = 1.$$

Then the average S is identical to the value $\varepsilon_{2k}^{(o)}$ obtained by the ε-algorithm since the preceding system is identical with the algebraic interpretation of the ε-algorithm given in [13, p. 51]. If $N > k$ we obtain extrapolation in the least squares sense by the ε-algorithm as described in [20] and [49]. More generally one can consider

$$\varepsilon_n = S - S_n + \sum_{i=1}^{k} a_i g_i(n)$$

$$I = \sum_{n=o}^{N} \varepsilon_n^2 \qquad\qquad N \geq k$$

and then find $a_1,...,a_k$ minimizing I.

Some statistical techniques were already applied to the problem of convergence acceleration in [188] but the subject has to be developed. On the other hand convergence acceleration methods could have some interesting applications in statistical procedures such as Monte-Carlo methods, a subject never studied as far as I know.

APPENDIX 1

A direct proof of the Christoffel-Darboux identity and a consequence.

For the usual orthogonal polynomials, the Christoffel-Darboux identity is always proved by using the three-terms recurrence relationship. We shall now give a sketch of a direct proof of this identity. For the details the reader is referred to [35].

We have

$$P_{k}(x) = t_k \begin{vmatrix} c_0 & ... & c_k \\ c_1 & ... & c_{k+1} \\ ... & ... & ... \\ c_{k-1} & ... & c_{2k-1} \\ 1 & ... & x^k \end{vmatrix} / G_k$$

where $G_k = \begin{vmatrix} c_0 & ... & c_{k-1} \\ c_1 & ... & c_k \\ ... & ... & ... \\ c_{k-1} & ... & c_{2k-2} \end{vmatrix}$ and where t_k is a non zero constant. Thus

$$P_{k}(x) = t_k x^k + \text{lower terms}.$$

We set $h_k = c(P_k^2(x)) = t_k^2 \, G_{k+1}/G_k$.
Let us define $K_k(x,t)$ by

$$K_{k}(x,t) \, G_{k+1} = - \begin{vmatrix} c_0 & c_1 & ... & c_k & 1 \\ c_1 & c_2 & ... & c_{k+1} & t \\ ... & ... & ... & ... & ... \\ c_k & c_{k+1} & ... & c_{2k} & t^k \\ 1 & x & ... & x^k & 0 \end{vmatrix} = - \begin{vmatrix} 0 & 1 & ... & x^k \\ 1 & c_0 & ... & c_k \\ t & c_1 & ... & c_{k+1} \\ ... & ... & ... & ... \\ t^k & c_k & .... & c_{2k} \end{vmatrix} .$$

Applying Sylvester's identity (see appendix 3) we obtain

$$\begin{vmatrix} 0 & 1 & ... & x^k \\ 1 & c_0 & ... & c_k \\ ... & ... & ... & ... \\ t^k & c_k & .... & c_{2k} \end{vmatrix} G_k = \begin{vmatrix} 0 & 1 & ... & x^{k-1} \\ 1 & c_0 & ... & c_{k-1} \\ ... & ... & ... & ... \\ t^{k-1} & c_{k-1} & .... & c_{2k-2} \end{vmatrix} G_{k+1}$$

$$- \begin{vmatrix} 1 & \ldots & x^k \\ c_0 & \ldots & c_k \\ \ldots & \ldots & \ldots \\ c_{k-1} & \ldots & c_{2k-1} \end{vmatrix} \begin{vmatrix} 1 & c_0 & \ldots & c_{k-1} \\ \ldots & \ldots & \ldots & \ldots \\ t^k & c_k & \ldots & c_{2k-1} \end{vmatrix}.$$

That is

$$- G_k G_{k+1} K_k(x,t) = -G_k G_{k+1} K_{k-1}(x,t) - (-1)^k G_k \frac{P_k(x)}{t_k} (-1)^k G_k \frac{P_k(t)}{t_k}$$

or

$$K_k(x,t) = K_{k-1}(x,t) + \frac{G_k^2}{t_k^2 G_k G_{k+1}} P_k(x) P_k(t) = K_{k-1}(x,t) + h_k^{-1} P_k(x) P_k(t)$$

and thus we obtain the known formula

$$K_k(x,t) = \sum_{i=0}^{k} h_i^{-1} P_i(x) P_i(t).$$

Let us now apply Schweins' formula (see appendix 3) to

$$\begin{vmatrix} 1 & \ldots & x^{k+1} \\ 1 & \ldots & t^{k+1} \\ c_0 & \ldots & c_{k+1} \\ \ldots & \ldots & \ldots \\ c_{k-1} & \ldots & c_{2k} \end{vmatrix}.$$

We have

$$\begin{vmatrix} 1 & \ldots & x^{k+1} \\ 1 & \ldots & t^{k+1} \\ c_0 & \ldots & c_{k+1} \\ \ldots & \ldots & \ldots \\ c_{k-1} & \ldots & c_{2k} \end{vmatrix} G_{k+1} = \begin{vmatrix} 1 & \ldots & x^{k+1} \\ c_0 & \ldots & c_{k+1} \\ \ldots & \ldots & \ldots \\ c_k & \ldots & c_{2k+1} \end{vmatrix} \begin{vmatrix} 1 & \ldots & t^k \\ c_0 & \ldots & c_k \\ \ldots & \ldots & \ldots \\ c_{k-1} & \ldots & c_{2k-1} \end{vmatrix}$$

$$- \begin{vmatrix} 1 & \ldots & x^k \\ c_0 & \ldots & c_k \\ \ldots & \ldots & \ldots \\ c_{k-1} & \ldots & c_{2k-1} \end{vmatrix} \begin{vmatrix} 1 & \ldots & t^{k+1} \\ c_0 & \ldots & c_{k+1} \\ \ldots & \ldots & \ldots \\ c_k & \ldots & c_{2k+1} \end{vmatrix}$$

$$= (-1)^{k+1} \frac{G_{k+1}}{t_{k+1}} P_{k+1}(x) (-1)^k \frac{G_k}{t_k} P_k(t) - (-1)^k \frac{G_k}{t_k} P_k(x) (-1)^{k+1} \frac{G_{k+1}}{t_{k+1}} P_{k+1}(t)$$

$$= - \frac{G_k G_{k+1}}{t_k t_{k+1}} \left[P_{k+1}(x) \, P_k(t) - P_k(x) \, P_{k+1}(t) \right].$$

Thus we finally have

$$
\begin{vmatrix}
1 & \dots & x^{k+1} \\
1 & \dots & t^{k+1} \\
c_0 & \dots & c_{k+1} \\
\dots & \dots & \dots \\
c_{k-1} & \dots & c_{2k}
\end{vmatrix}
= - \frac{G_k}{t_k t_{k+1}} \left[P_{k+1}(x) P_k(t) - P_k(x) P_{k+1}(t) \right]
$$

$$
= - \frac{t_k}{t_{k+1} h_k} G_{k+1} \left[P_{k+1}(x) P_k(t) - P_k(x) P_{k+1}(t) \right].
$$

We shall now prove that

$$
(x-t)
\begin{vmatrix}
0 & 1 & \dots & x^k \\
1 & c_0 & \dots & c_k \\
\dots & \dots & \dots & \dots \\
t^k & c_k & \dots & c_{2k}
\end{vmatrix}
=
\begin{vmatrix}
1 & \dots & x^{k+1} \\
1 & \dots & t^{k+1} \\
c_0 & \dots & c_{k+1} \\
\dots & \dots & \dots \\
c_{k-1} & \dots & c_{2k}
\end{vmatrix}
\tag{*}.
$$

In [35], three different proofs of this identity are given. The first one involves the reproducing property of $K_k(x,t)$ that is $\forall p \in P_k$, $c(p(x) \, K_k(x,t)) = p(t)$ and the fact that the functionals $L_i(.) = c(x^i.)$ are linearly independent since $G_{k+1} \neq 0$. The second proof is due to Prévost [158] ; it is by recurrence and uses Sylvester's identity. The last proof, which is the simplest one, was obtained by Hendriksen [95]. It is as follows. Taking the determinant in the right hand side of (*) we replace each column (from the second one) by its difference with the preceding one multiplied by x and then we put t-x in factor. We obtain

$$
\begin{vmatrix}
1 & x & \dots & x^{k+1} \\
1 & t & \dots & t^{k+1} \\
c_0 & c_1 & \dots & c_{k+1} \\
\dots & \dots & \dots & \dots \\
c_{k-1} & c_k & \dots & c_{2k}
\end{vmatrix}
=
\begin{vmatrix}
1 & 0 & \dots & 0 \\
1 & t-x & \dots & t^{k+1} - x t^k \\
c_0 & c_1 - x c_0 & \dots & c_{k+1} - x c_k \\
\dots & \dots & \dots & \dots \\
c_{k-1} & c_k - x c_{k-1} & \dots & c_{2k} - x c_{2k-1}
\end{vmatrix}
$$

$$
= (t-x)
\begin{vmatrix}
1 & \dots & t^k \\
c_1 - x c_0 & \dots & c_{k+1} - x c_k \\
\dots & \dots & \dots \\
c_k - x c_{k-1} & \dots & c_{2k} - x c_{2k-1}
\end{vmatrix}.
$$

Then we add a new second row $(1, c_0,...,c_k)$, a new first column $(0, 1, 0,...,0)$ and we change the sign. Finally we multiply each row (from the second one) by x and we add to the following one. Thus we get

$$(x-t) \begin{vmatrix} 0 & 1 & ... & t^k \\ 1 & c_0 & ... & c_k \\ 0 & c_1-xc_0 & ... & c_{k+1}-xc_k \\ ... & ... & ... & ... \\ 0 & c_k-xc_{k-1} & ... & c_{2k}-xc_{2k-1} \end{vmatrix} = (x-t) \begin{vmatrix} 0 & 1 & ... & t^k \\ 1 & c_0 & ... & c_k \\ x & c_1 & ... & c_{k+1} \\ ... & ... & ... & ... \\ x^k & c_k & ... & c_{2k} \end{vmatrix}$$

and (*) is proved.

Thus we have

$$-\frac{t_k}{t_{k+1}h_k} G_{k+1} \left[P_{k+1}(x) P_k(t) - P_k(x) P_{k+1}(t) \right] = -(x-t) G_{k+1} K_k(x,t)$$

and we finally obtain the usual Christoffel-Darboux identity

$$\frac{t_k}{t_{k+1}h_k} \left[P_{k+1}(x) P_k(t) - P_k(x) P_{k+1}(t) \right] = (x-t) \sum_{i=0}^{k} h_i^{-1} P_i(x) P_i(t).$$

Now we can ask the question whether a family of polynomials satisfying the Christoffel-Darboux identity also satisfies a three-terms recurrence relationship. Thus let $\{P_k\}$ be a family of polynomials (which are not assumed to be orthogonal) such that $\forall k \geq 0$

- P_k has the exact degree k

- $\gamma_k[P_{k+1}(x) P_k(t) - P_{k+1}(t) P_k(x)] = (x-t) \sum_{i=0}^{k} a_i P_i(x) P_i(t)$ (**)

where the a_i's are constants independent of k and γ_k is a non zero constant.

We have

$$\gamma_k \left[P_{k+1}(x) P_k(t) - P_{k+1}(t) P_k(x) \right] = (x-t) a_k P_k(x) P_k(t) + (x-t) \sum_{i=0}^{k-1} a_i P_i(x) P_i(t)$$

$$= (x-t) a_k P_k(x) P_k(t) + (x-t) \gamma_{k-1} \left[P_k(x) P_{k-1}(t) - P_k(t) P_{k-1}(x) \right].$$

Thus, $\forall x, t$

$$P_k(t)\left[\gamma_k P_{k+1}(x) - a_k x P_k(x) + \gamma_{k-1} P_{k-1}(x)\right]$$
$$= P_k(x)\left[\gamma_k P_{k+1}(t) - a_k t P_k(t) + \gamma_{k-1} P_{k-1}(t)\right].$$

That is

$$\left[\gamma_k P_{k+1}(x) - a_k x P_k(x) + \gamma_{k-1} P_{k-1}(x)\right] / P_k(x) = b_k$$

where b_k is a constant independent of x.

This is equivalent to

$$\gamma_k P_{k+1}(x) = (a_k x + b_k) P_k(x) - \gamma_{k-1} P_{k-1}(x) \qquad (***)$$

which shows that if the Christoffel-Darboux identity holds then the polynomials $\{P_k\}$ satisfy a 3-terms recurrence relationship that is, by an extension due to Shohat [172] of a theorem by Favard [68], they form a family of formal orthogonal polynomials with respect to a linear functional c whose moments can be calculated, see [17, p. 155] and [162].

Let us now find the expressions of the constants γ_k, a_k and b_k. It is easy to see that if we write P_k as $P_k(x) = t_k x^k + $ lower terms, then $\gamma_k t_{k+1} = a_k t_k$. We set $A_{k+1} = a_k / \gamma_k$. Multiplying $(***)$ by x^{k-1} and applying c gives

$$\gamma_k c(x^{k-1} P_{k+1}(x)) - a_k c(x^k P_k(x)) - b_k\, c\, (x^k P_k(x)) + \gamma_{k-1}\, c(x^{k-1} P_{k-1}(x)) = 0$$

or

$$a_k\, c(x^k P_k(x)) = \gamma_{k-1}\, c(x^{k-1} P_{k-1}(x)).$$

But $h_k = t_k c(x^k P_k(x))$ and we have

$$\frac{a_k h_k}{t_k} = \gamma_{k-1} \frac{h_{k-1}}{t_{k-1}}$$

since $\forall k$, $t_k \neq 0$.

Thus, setting $C_{k+1} = \gamma_{k-1}/\gamma_k$, we have

$$C_{k+1} = \frac{a_k h_k t_{k-1}}{t_k h_{k-1} \gamma_k} = \frac{t_{k+1}}{t_k} \frac{h_k t_{k-1}}{t_k h_{k-1}} = \frac{t_{k-1} t_{k+1}}{t_k^2} \frac{h_k}{h_{k-1}}.$$

Multiplying (∗∗∗) by P_k and applying c, we get

$$\gamma_k \, c \, (P_k(x) \, P_{k+1}(x)) - a_k c(x \, P_k^2 (x)) - b_k c(P_k^2(x)) + \gamma_{k-1} \, c(P_k(x) \, P_{k-1}(x)) = 0$$

that is

$$b_k = - c(x \, P_k^2(x))/h_k.$$

Setting $B_{k+1} = b_k/\gamma_k$ and $\alpha_k = c(x \, P_k^2(x))$ we have

$$B_{k+1} = - \frac{\alpha_k a_k}{h_k \gamma_k} = - \frac{\alpha_k}{h_k} \frac{t_{k+1}}{tk} \, .$$

Thus we have finally proved that

$$P_{k+1}(x) = (A_{k+1}x + B_{k+1}) \, P_k(x) - C_{k+1} \, P_{k-1}(x)$$

with

$$A_{k+1} = \, t_{k+1}/t_k, \quad B_{k+1} = - \frac{\alpha_k}{h_k} \frac{t_{k+1}}{t_k} \text{ and } C_{k+1} = \frac{t_{k-1}t_{k+1}}{t_k^2} \frac{h_k}{h_{k-1}}$$

which is the usual recurrence relationship.
Moreover

$$A_{k+1} = \frac{a_k}{\gamma_k} = \frac{t_{k+1}}{t_k} \text{ and } C_{k+1} = \frac{\gamma_{k-1}}{\gamma_k} = \frac{A_{k+1}h_k}{A_k h_{k-1}}$$

and thus

$$C_{k+1} = \frac{a_k}{\gamma_k} \frac{h_k}{h_{k-1}} \frac{\gamma_{k-1}}{a_{k-1}} = \frac{a_k h_k}{a_{k-1}h_{k-1}} C_{k+1}.$$

It follows that $\exists \gamma \neq 0$ such that $\forall k$, $a_k h_k = \gamma$ or $a_k = \gamma h_k^{-1}$.
Thus

$$\gamma_k = a_k \frac{t_k}{t_{k+1}} = \, \gamma \frac{t_k}{h_k t_{k+1}}$$

and (∗∗) becomes

$$\frac{t_k}{\bar{h}_k t_{k+1}}\left[P_{k+1}(x)\,P_k(t) - P_{k+1}(t)\,P_k(x)\,\right] = (x\text{-}t) \sum_{i=o}^{k} h_i^{-1}\,P_i(x)\,P_i(t)$$

which shows the equivalence between the Christoffel-Darboux identity and the three-terms recurrence relationship.

Let us set

$$(P_{k+1}(x)P_k(t) - P_{k+1}(t)\,P_k\,(x))/(x\text{-}t) = \sum_{i,j=1}^{k+1} a_{ij}x^{\,i\text{-}1}\,t^{\,j\text{-}1}.$$

This polynomial in two variables is related to the determinants of relation (*). The matrix $A = (a_{ij})$ is the so-called Bezoutian matrix of the polynomials P_k and P_{k+1}. Let us recall that its inverse (which exists if and only if P_k and P_{k+1} have no common zero) is a Hankel matrix [4] and conversely. Bezoutian matrices, whose properties can be found in [69], have many applications in linear control systems, electrical networks, signal processing, and coding theory [7].

APPENDIX 2.

Duality in Padé-type approximation.

Let V_k be an arbitrary polynomial of degree k and let R_k be the Hermite interpolation polynomial of $(1-xt)^{-1}$ at the zeros of V_k. Then $c(R_k)$ is the so-called Padé-type approximant of f with generating polynomial V_k. It is a rational function with a numerator of degree k-1 and a denominator of degree k, denoted by $(k-1/k)_f(t)$ and such that

$$f(t) - (k-1/k)_f(t) = O(t^k) \qquad (t \to 0).$$

If V_k is identical to the formal orthogonal polynomial P_k with respect to c, that is the polynomial satisfying the orthogonality conditions

$$c(x^i P_k(x)) = 0 \qquad i = 0,...,k-1$$

then the Padé-type approximant $(k-1/k)_f(t)$ becomes identical to the classical Padé approximant $[k-1/k]_f(t)$ such that

$$f(t) - [k-1/k]_f(t) = O(t^{2k}) \qquad (t \to 0).$$

The aim of this appendix is to give some properties of the functional d (depending on V_k) such that

$$d((1-xt)^{-1}) = c(R_k) = (k-1/k)_f(t).$$

For conveniency reasons, we shall make use of the notation of duality

$$\langle L, g \rangle$$

to denote the action of the linear functional L on the element g of a vector space E. Thus L belongs to E^*, the dual space of E, that is the vector space of linear functionals on E. If T is a linear operator mapping E into itself, the dual operator T^* of T is the linear operator mapping E^* into itself, which is uniquely defined by

$$\langle T^*(L), g \rangle = \langle L, Tg \rangle$$

$\forall L \in E^*$ and $\forall g \in E$, [161].

Now, let E be the space of functions which are holomorphic in a neighbourhood of the origin and let V_k be an arbitrary polynomial of degree k, with distincts zeros $x_1,...,x_n$ of respective multiplicities $k_1,...,k_n$ and $k_1+...+k_n=k$.

Let $I(V_k)$ be the linear operator mapping $g \in E$ into its Hermite interpolation polynomial R_k of degree at most k-1 defined by

$$g^{(j)}(x_i) = R_k^{(j)}(x_i) \quad \text{for } i = 1,...,n \text{ and } j = 0,...,k_i\text{-}1.$$

Let $\tilde{V}_k(t) = t^k V_k(t^{-1})$ and let U_k be the reciprocal series of $\tilde{V}_k$ (which exists since $\tilde{V}_k(0) \neq 0$) formally defined by

$$U_k(t)\ \tilde{V}_k(t) = 1.$$

We set

$$V_k(x) = v_0 + v_1 x + ... + v_k x^k$$

$$U_k(t) = u_0 + u_1 t + u_2 t^2 + ...$$

Then

$$\tilde{V}_k(t) = v_0 t^k + v_1 t^{k-1} + ... + v_k$$

and we have

$$u_0 v_k = 1$$
$$u_0 v_{k-1} + u_1 v_k = 0$$
$$\text{------------------}$$
$$u_0 v_0 + u_1 v_1 + ... + u_k v_k = 0$$
$$u_1 v_0 + u_2 v_1 + ... + u_{k+1} v_k = 0$$
$$\text{----------------------------}$$

That is, with the convention that $u_j = 0$ for $j < 0$

$$u_0 v_k = 1$$

$$v_0 u_i + v_1 u_{i+1} + ... + v_k u_{i+k} = 0 \quad , \quad i \neq -k.$$

We set

$$r_i(x) = \sum_{j=0}^{i} u_j x^{i-j} \quad , \quad i \geq 0$$

$$r_i(x) = 0 \qquad\qquad , \quad i < 0.$$

Lemma 1. *For all* $i \geq 0$

$$x^i - r_{i-k}(x) \; V_k(x) = -\sum_{j=o}^{k-1} a_j^{(i)} \; x^j \quad with \quad a_j^{(i)} = \sum_{m=o}^{j} v_m \; u_{i-k+m-j}.$$

The proofs of the results will be omitted. They can be found in [36].

Lemma 2. *For all* $i \geq 0$
$$I(V_k) \; x^i = x^i - r_{i-k}(x) \; V_k(x).$$

Lemma 3. $\qquad\qquad I(V_k)(1-xt)^{-1} = (1-xt)^{-1} \; (1-t^k \; V_k(x)/\tilde{V}_k(t)).$

Lemma 3 thus provides a new proof of a known result.

As we saw above

$$(k-1/k)_f(t) = c(R_k(x)) = <c, \; I(V_k)(1-xt)^{-1}>.$$

Thus we have

$$(k-1/k)_f(t) = <I^*(V_k)(c), \; (1-xt)^{-1}>.$$

Let us set

$$d(V_k) = I^*(V_k)(c). \qquad\qquad (*)$$

We have

$$<d(V_k), \; x^i> = <c, \; I(V_k)x^i> = <c, \; x^i - r_{i-k}(x) \; V_k(x)> = d_i.$$

Since

$$(k-1/k)_f = <d(V_k), \; 1+xt+x^2t^2 + \ldots > = d_0 + d_1t + d_2t^2 + \ldots$$

then the operator which maps the formal power series f into the power series $(k-1/k)_f(t)$ can be understood as the mapping of E^* into itself which maps c into $d(V_k)$. This mapping, which depends on the generating polynomial V_k, will be called the Padé-type operator ; from

(*) we see that this operator is $I^*(V_k)$. If V_k does not depend on c then $I^*(V_k)$ is, as usual, linear. But for Padé approximats, since V_k is the orthogonal polynomial of degree k with respect to the functional c, then V_k depends on c and the linearity property only holds if the first 2k moments of both functionals are the same since, then, both orthogonal polynomials of degree k will be the same.

Let us now study some properties of $d(V_k)$.

We obviously have the

Property 1 :

$$<d(V_k), x^i> = <c, x^i> \quad \text{for } i = 0,...,k-1.$$

Moreover if $<c, x^i V_k(x)> = 0$ for i = 0,...,k-1 *then the preceding equality holds for* i = 0,...,2k-1.
In both cases, $\forall m \geq k$

$$<d(V_m), I(V_k)(1-xt)^{-1}> = <c, I(V_k)(1-xt)^{-1}>.$$

Property 2 : *For all* $i \geq 0$, $<d(V_k), x^i V_k(x)> = 0$.

This property, which is a generalization of a property given in [17, p. 23] when V_k has distinct simple zeros, can also be proved directly but the proof is much longer.

As a corollary of property 2 we get a recursive formula for computing the d_i's.

Corollary 1. *For* $i \geq 0$, *we have*

$$d_{i+k} = - (v_0 d_i + ... + v_{k-1} d_{i+k-1})/v_k$$

with $d_i = c_i$ for i = 0,...,k-1.

Thus, for given coefficients of V_k, the computation of all the d_i's only uses $c_0,...,c_{k-1}$. If V_k is the orthogonal polynomial of degree k with respect to c, then the computation of V_k needs the knowledge of $c_0,...,c_{2k-1}$.

The d_i's can be used as approximations of the missing c_i's, an idea introduced in [80] (see also [25]), and we immediately have an expression for the error.

Property 3 : *For all* $i \geq 0$

$$c_i - d_i = <c, r_{i-k}(x) \, V_k(x)>$$

with

$$<c, r_{i-k}(x) \, V_k(x)> = c_i + \sum_{j=0}^{k-1} a_j^{(i)} \, c_j$$

where the $a_j^{(i)}$ *can be recursively computed by*

$$a_o^{(i+1)} = u_{i-k+1} v_o$$

$$a_j^{(i+1)} = a_{j-1}^{(i)} + u_{i-k+1} v_j \qquad for \;\; j = 1,...,k-1$$

with

$$a_j^{(k-1)} = 0 \qquad\qquad j = 0,...,k-1.$$

As always in numerical analysis, this formula cannot be used in practice to compute the error $c_i - d_i$ since its computation needs the knowledge of the unkown coefficient c_i. However it can be useful in some cases. For example if

$$c(x^i) = \int_a^b x^i \, \alpha(x) dx \qquad\qquad i \geq 0$$

where α is positive in [a,b], then $\exists \beta \in$ [a,b] such that

$$c_i - d_i = c_o \, r_{i-k}(\beta) \, V_k(\beta)$$

and bounds for $c_i - d_i$ can be obtained.

Let us now consider a series of functions of the form

$$f(t) = \sum_{i=o}^{\infty} c_i g_i(t).$$

Let G be the generating function of the g_i's defined by

$$G(x,t) = \sum_{i=0}^{\infty} x^i g_i(t).$$

As above we formally have

$$f(t) = c(G(x,t)).$$

Let V_k be an arbitrary polynomial of degree k and let R_k be the Hermite interpolation polynomial of $G(.,t)$ at the zeros of V_k. We shall define the Padé-type approximant $(k-1/k)_f(t)$ of f as

$$(k-1/k)_f(t) = c(R_k(x)).$$

Usually $(k-1/k)_f(t)$ is not any more a rational function but we still have

$$f(t) - (k-1/k)_f(t) = O(g_k(t))$$

which means that $f(t) - (k-1/k)_f(t) = \sum_{i=k}^{\infty} d_i g_i(t).$

Let L be a linear functional transformation. We set

$$h_i(p) = Lg_i(t).$$

For example

$$h_i(p) = \int_0^{\infty} e^{-pt} g_i(t)dt.$$

If we set

$$F(p) = Lf(t)$$

it was proved in [37] that

$$(k-1/k)_F(p) = L(k-1/k)_f(t)$$

if both approximants have the same generating polynomial V_k (which is true in the Padé case since the functional c remains unchanged).

If $g_i(t) = t^i$ then we have

$$I(V_k) \ L(1-xt)^{-1} = L(1-xt)^{-1}(1-t^k V_k(x)/\tilde{V}_k(t))$$

and thus

$$(k-1/k)_F(p) = <I^*(V_k)(c), \ L(1-xt)^{-1}> = d_0 h_0(p) + d_1 h_1(p) + d_1 h_1(p) + \ldots$$

with the same d_i's as before

$$d_i = <d(V_k), \ x^i> = <c, \ x^i - r_{i-k}(x) \ V_k(x)>.$$

This result gives another justification of the definition used by van Rossum [163] for Padé approximants to series of functions.

Since $(k-1/k)_F(p)$ approximates $F(p)$, $L^{-1}(k-1/k)_F(p)$ approximates $L^{-1}F(p) = f(t)$. But

$$L^{-1}(k-1/k)_F(p) = (k-1/k)_f(t).$$

Thus if the expansion of $(k-1/k)_F(p)$ is known, that of $(k-1/k)_f(t)$ is obtained by replacing the $h_i(p)$ by the $g_i(t)$. This was the method used by Longman for inverting the Laplace transform by means of Padé approximants [128] or by Brezinski by means of Padé-type approximants [16]. In these cases the summation of the infinite series can be avoided by a special trick due to Longman and Sharir [129]. The convergence was studied by van Iseghem [104] (see also [37]).

APPENDIX 3

Sylvester's and Schweins' identities in a vector space.

Let $b_1,...,b_n$ be elements of a vector space and let a_{ij} be a scalar, $\forall$ i and j. Then, Sylvester's identity is

$$\begin{vmatrix} b_1 & ... & b_n \\ a_{11} & ... & a_{1n} \\ ... & ... & ... \\ a_{n-1,1}...a_{n-1,n} \end{vmatrix} \begin{vmatrix} a_{12} & ... & a_{1,n-1} \\ ... & ... & ... \\ a_{n-2,2}...a_{n-2,n-1} \end{vmatrix} = \begin{vmatrix} b_1 & ... & b_{n-1} \\ a_{11} & ... & a_{1,n-1} \\ ... & ... & ... \\ a_{n-2,1}...a_{n-2,n-1} \end{vmatrix} \begin{vmatrix} a_{12} & ...a_{1n} \\ ... & ... & ... \\ a_{n-1,2}...a_{n-1} \end{vmatrix}$$

$$- \begin{vmatrix} b_2 & ... & b_n \\ a_{12} & ... & a_{1n} \\ ... & ... & ... \\ a_{n-2,2}...a_{n-2,n} \end{vmatrix} \begin{vmatrix} a_{11} & ... & a_{1,n-1} \\ ... & ... & ... \\ a_{n-1,1}...a_{n-1,n-1} \end{vmatrix} .$$

Let now $c_1,...,c_n$ be scalars. Then Schweins' identity is

$$\begin{vmatrix} b_1 & ... & b_n \\ a_{11} & ... & a_{1n} \\ ... & ... & ... \\ a_{n-1,1}...a_{n-1,n} \end{vmatrix} \begin{vmatrix} c_1 & ... & c_{n-1} \\ a_{11} & ... & a_{1,n-1} \\ ... & ... & ... \\ a_{n-2,1}...a_{n-2,n-1} \end{vmatrix} - \begin{vmatrix} b_1 & ... & b_{n-1} \\ a_{11} & ... & a_{1,n-1} \\ ... & ... & ... \\ a_{n-2,1}...a_{n-2,n-1} \end{vmatrix} \begin{vmatrix} c_1 & ... & c_n \\ a_{11} & ... & a_{1n} \\ ... & ... & ... \\ a_{n-1,1}...a_{n-1,n} \end{vmatrix}$$

$$= \begin{vmatrix} b_1 & ... & b_n \\ c_1 & ... & c_n \\ a_{11} & ... & a_{1n} \\ ... & ... & ... \\ a_{n-2,1}...a_{n-2,n} \end{vmatrix} \begin{vmatrix} a_{11} & ... & a_{1,n-1} \\ ... & ... & ... \\ a_{n-1,1}...a_{n-1,n-1} \end{vmatrix} .$$

If $c_1 = 1$ and $c_2 = ... = c_n = 0$, then Schweins' identity reduces to Sylvester's.
Schweins' identity also holds if $b_1,...,b_n$ are scalars and $c_1,...,c_n$ elements of a vector space. It reduces to Sylvester's if $b_1 = 1$, $b_2 = ... = b_n = 0$. For a proof of these identities see [23].
Let me give a quite simple proof of Sylvester's identity in the scalar case which only requires to know that the determinant of a block triangular matrix is equal to the product of the determinants of the blocks on its diagonal.

α, β, γ, δ are scalars
A is a square matrix n x n
a, c are row vectors of dimension n
b,d are column vectors of dimension n.

We shall compute, by two different ways, the determinant

$$\begin{vmatrix} \alpha & a & 0 & \beta \\ b & A & 0 & d \\ b & 0 & A & d \\ \gamma & 0 & c & \delta \end{vmatrix}.$$

Replacing the third row by its difference with the second one, we get

$$\begin{vmatrix} \alpha & a & 0 & \beta \\ b & A & 0 & d \\ 0 & -A & A & 0 \\ \gamma & 0 & c & \delta \end{vmatrix}.$$

Replacing the second column by its sum with the third one leads to

$$\begin{vmatrix} \alpha & a & 0 & \beta \\ b & A & 0 & d \\ 0 & 0 & A & 0 \\ \gamma & c & c & \delta \end{vmatrix} = \begin{vmatrix} \alpha & a & \beta & 0 \\ b & A & d & 0 \\ \gamma & c & \delta & c \\ 0 & 0 & 0 & A \end{vmatrix} = |A| \begin{vmatrix} \alpha & a & \beta \\ b & A & d \\ \gamma & c & \delta \end{vmatrix}.$$

The second method for computing the initial determinant is as follows. It is equal to the sum

$$\begin{vmatrix} \alpha & a & 0 & \beta \\ b & A & 0 & d \\ 0 & 0 & A & d \\ 0 & 0 & c & \delta \end{vmatrix} + \begin{vmatrix} 0 & a & 0 & \beta \\ 0 & A & 0 & d \\ b & 0 & A & d \\ \gamma & 0 & c & \delta \end{vmatrix} =$$

$$\begin{vmatrix} \alpha & a & 0 & \beta \\ b & A & 0 & d \\ 0 & 0 & A & d \\ 0 & 0 & c & \delta \end{vmatrix} + (-1)^n (-1)^{n+1} \begin{vmatrix} a & \beta & 0 & 0 \\ A & d & 0 & 0 \\ 0 & d & b & A \\ 0 & \delta & \gamma & c \end{vmatrix} =$$

$$\begin{vmatrix} \alpha & a \\ b & A \end{vmatrix} \begin{vmatrix} A & d \\ c & \delta \end{vmatrix} - \begin{vmatrix} a & \beta \\ A & d \end{vmatrix} \begin{vmatrix} b & A \\ \gamma & c \end{vmatrix}$$

which ends the proof.

In the scalar case, Schweins' identity can be immediately obtained by applying Sylvester's to

$$
\begin{vmatrix}
0 & b_1 & \dots & b_{n-1} & b_n \\
0 & a_{11} & \dots & a_{1,n-1} & a_{1n} \\
\dots & \dots & \dots & \dots & \dots \\
0 & a_{n-2,1} & \dots & a_{n-2,n-1} & a_{n-2,n} \\
1 & a_{n-1,1} & \dots & a_{n-1,n-1} & a_{n-1,n} \\
0 & c_1 & \dots & c_{n-1} & c_n
\end{vmatrix}
$$

For determinantal identities, see [3].

References

[1] J. ABAFFY, C. BROYDEN, E. SPEDICATO
A class of direct methods for linear systems.
Numer. Math., 45 (1984), 361-376.

[2] S. ACHAKIR
Connexion entre les méthodes de point fixe et d'accélération de la convergence.
Thèse de 3ème cycle, Université de Lille 1, 1982.

[3] A.C. AITKEN
Determinants and matrices.
Oliver and Boyd, Edinburgh, 1951.

[4] M. ALVAREZ, G. SANSIGRE
On polynomials with interlacing zeros.
in "Polynômes Orthogonaux et Applications", C. Brezinski et al. eds., LNM 1171, Springer-Verlag, Berlin, 1985, pp. 255-258.

[5] G.A. BAKER Jr.
Application of the Padé approximant method to the investigation of some magnetic properties of the Ising model.
Phys. Rev., 124 (1961) 768-774.

[6] S. BANACH
Théorie des opérations linéaires.
Monografie Matematyczne vol. 1, Warszawa, 1932.

[7] S. BARNETT
Polynomials and linear control systems.
Marcel Dekker, New-York, 1983.

[8] A. BELANTARI
Procédures d'estimation de l'erreur dans l'approximation de type-Padé.
Thèse, Université de Lille 1, 1989.

[9] A. BERLINET
Sequence transformations as statistical tools.
Appl. Numer. Math, 1 (1985), 531-544.

[10] N.M. BLACHMAN
Random processes and orthogonal linear forms.
Preprint, 1989.

[11] C. BREZINSKI
Etudes sur les ε et ρ-algorithmes.
Numer. Math., 17 (1971), 153-162.

[12] C. BREZINSKI
Généralisations de la transformation de Shanks, de la table de Padé et de l' ε-algorithme.
Calcolo, 12 (1975), 317-360.

[13] C. BREZINSKI
Accélération de la convergence en analyse numérique.
LNM 584, Springer-Verlag, Berlin, 1977.

[14] C. BREZINSKI
Algorithmes d'accélération de la convergence. Etude numérique.
Editions Technip, Paris, 1978.

[15] C. BREZINSKI
Sur le calcul de certains rapports de déterminants.
in *"Padé Approximation and its Applications"*, L. Wuytack ed., LNM 765, Springer-Verlag, Berlin, 1979, pp. 184-210.

[16] C. BREZINSKI
Rational approximation to formal power series.
J. Approximation Theory, 25 (1979), 295-317.

[17] C. BREZINSKI
Padé-type approximation and general orthogonal polynomials.
ISNM vol. 50, Birkhäuser-Verlag, Basel, 1980.

[18] C. BREZINSKI
A general extrapolation algorithm.
Numer. Math., 35 (1980), 175-187.

[19] C. BREZINSKI
The Mühlbach-Neville-Aitken algorithm and some extensions.
BIT, 20 (1980), 444-451.

[20] C. BREZINSKI
Algorithm 585 : A subroutine for the general interpolation and extrapolation problems.
Trans. on Math. Soft., 8 (1982), 290-301.

[21] C. BREZINSKI
Recursive interpolation, extrapolation and projection.
J. Comput. Appl. Math., 9 (1983), 369-376.

[22] C. BREZINSKI
About Henrici's method for nonlinear equations.
Symposium on Numerical Analysis and Computational Complex
Analysis, Zürich, August 15-17, 1983, unpublished.

[23] C. BREZINSKI
Some determinantal identities in a vector space, with applications.
in *"Padé Approximation and its Applications - Bad-Honnef 1983"* H
Werner and H.J. Bünger eds., LNM 1071, Springer-Verlag, Berlin,
1984, pp. 1-11.

[24] C. BREZINSKI
Composite sequence transformations.
Numer. Math., 46 (1985), 311-321.

[25] C. BREZINSKI
Prediction properties of some extrapolation methods.
Appl. Numer. Math., 1 (1985), 457-462.

[26] C. BREZINSKI
*Ideas for further investigations on orthogonal polynomials and Padé
approximants.*
Actas III Simposium sobre Polinomios Ortogonales y Aplicaciones, F.
Marcellan ed., 1985.

[27] C. BREZINSKI
Error estimate in Padé approximation.
in *"Orthogonal Polynomials and their Applications"*, M. Alfaro et al.
eds., LNM 1329, Springer-Verlag, Berlin, 1988, pp. 1-19.

[28] C. BREZINSKI
Other manifestations of the Schur complement.
Linear Algebra. Appl., 111 (1988), 231-247.

[29] C. BREZINSKI
Quasi-linear extrapolation processes.
in *"Numerical Mathematics. Singapore 1988"*, R.P. Agarwal et al.
eds., ISNM vol. 86, Birkhäuser-Verlag, Basel, 1988, pp. 61-78.

[30] C. BREZINSKI
Partial Padé approximation.
J. Approximation Theory, 54 (1988), 210-233.

[31] C. BREZINSKI
Algebraic properties of the E-transformation.
in *"Numerical analysis and mathematical modelling""*, Banach Center Publications, vol. 24, PWN, Warsaw, 1990, pp. 85-90.

[32] C. BREZINSKI
Bordering methods and progressive forms for sequence transformations.
Zastosow. Mat., 20 (1990), 435-443.

[33] C. BREZINSKI
A survey of iterative extrapolation by the E-algorithm.
Det Kong. Norske Vid. Selsk. Skr., 32 (1990) 17-25

[34] C. BREZINSKI
Orthogonal and Stieltjes polynomials with respect to an even functional.
Rend. Sem. Mat. Univ. Torino, 45 (1987), 75-82.

[35] C. BREZINSKI
A direct proof of the Christoffel-Darboux identity and its equivalence to the recurrence relationship.
J. Comput. Appl. Math., 32 (1990), 17-25.

[36] C. BREZINSKI
Duality in Padé-type approximation.
J. Comput. Appl. Math., 30 (1990), 351-357.

[37] C. BREZINSKI, J. VAN ISEGHEM
Padé-type approximants and linear functional transformations.
in *"Rational Approximation and Interpolation"*, P.R. Graves-Morris et al. eds., LNM 1105, Springer-Verlag, Berlin, 1984, pp. 100-108.

[38] C. BREZINSKI, J. VAN ISEGHEM
Padé approximations.
Handbook of Numerical Analysis, North-Holland, Amsterdam, to appear.

[39] C. BREZINSKI, M. REDIVO ZAGLIA
Extrapolation methods. Theory and practice.
North-Holland, Amsterdam, to appear.

[40] C. BREZINSKI, H. SADOK
Vector sequence transformations and fixed point methods.
in *"Numerical Methods in Laminar and Turbulent Flows"*, C. Taylor et al. eds., Pineridge Press, Swansea, 1987, pp. 3-11.

[41] M.G. DE BRUIN
Simultaneous Padé approximation and orthogonality.
in *"Polynômes Orthogonaux et Applications"*, C. Brezinski et al. eds., LNM 1171, Springer-Verlag, Berlin, 1985, pp. 74-83.

[42] M.G. DE BRUIN
Simultaneous partial Padé approximants.
J. Comput. Appl. Math., 21 (1988), 343-355.

[43] A. BULTHEEL
Laurent series and their Padé approximations.
Birkhäuser-Verlag, Basel, 1987.

[44] S.K. BURLEY, S.O. JOHN, J. NUTTALL
Vector orthogonal polynomials.
SIAM J. Numer. Anal., 18 (1981), 919-924.

[45] S. CABAY, L.W. JACKSON
A polynomial extrapolation method for finding limits and antilimits of vector sequences.
SIAM J. Numer. Anal., 13 (1976), 734-752.

[46] C. CARSTENSEN
On a general ε-algorithm.
in *" Numerical and Applied Mathematics"*, W.F. Ames and C. Brezinski eds., Baltzer, Basel, 1989, vol. 1.2, pp. 437-441.

[47] J.P. COLEMAN
Numerical methods for $y'' = f(x,y)$ via rational approximations for the cosine.
IMA J. Numer. Anal., 9 (1989), 145-165.

[48] F. CORDELLIER
Utilisation de l'invariance homographique dans les algorithmes de losange.
in *"Padé Approximation and its Applications. Bad-Honnef 1983"*, H. Werner and H.J. Bünger eds., LNM 1071, Springer-Verlag, Berlin, 1984, pp. 62-94.

[49] F. CORDELLIER
Interpolation rationnelle et autres questions. Problèmes algorithmiques et numériques.
Thèse d'Etat, Université de Lille 1, 1989.

[50] C.W. CRYER
Numerical functional analysis.
Clarendon Press, Oxford, 1982.

[51] A. CUYT
General order multivariate rational Hermite interpolants.
Habilitation, University of Antwerp, 1986.

[52] A. CUYT
Multivariate Padé approximants revisited.
BIT, 26 (1986), 71-79.

[53] A. CUYT
A recursive computation scheme for multivariate rational interpolants.
SIAM J. Numer. Anal., 24 (1987), 228-239.

[54] A. CUYT
Old and new multidimensional convergence accelerators.
Appl. Numer. Math., 6 (1989/1990), 169-185.

[55] A. CUYT, B. VERDONK
Different techniques for the construction of multivariate rational interpolants.
in "Nonlinear Numerical Methods and Rational Approximation",
A. Cuyt ed., Reidel, Dordrecht, 1988, pp. 167-190.

[56] A. CUYT, L. WUYTACK
Nonlinear methods in numerical analysis.
North-Holland, Amsterdam, 1987.

[57] P.J. DAVIS
Interpolation and approximation.
Dover, New-York, 1975.

[58] G. DAHLQUIST, A. BJÖRCK
Numerical methods.
Prentice-Hall, Englewood Cliffs, 1974.

[59] J.P. DELAHAYE
Sequence transformations.
Springer-Verlag, Berlin, 1988.

[60] J. DELLA DORA
Padé-Hermite approximants.
in *"Numerical Methods in the Study of Critical Phenomena"*, J. Della Dora et al. eds., Springer-Verlag, Berlin, 1981, pp. 3-11.

[61] J. DERUYTS
Sur une classe de polynômes conjugués.
Mémoires couronnés et Mémoires des Savants Etrangers, Acad. Roy. Sci. Lett. Beaux Arts de Belgique, 48 (1886).

[62] M.F. DIDON
Sur certains systèmes de polynômes associés.
Ann. Sci. Ec. Norm. Sup., 6 (1869), 111-125.

[63] N.R. DRAPER, H. SMITH
Applied regression analysis.
J. Wiley, New York, 1966.

[64] A. DRAUX
Polynômes orthogonaux formels. Applications.
LNM 974, Springer-Verlag, Berlin, 1983.

[65] A. DRAUX, A. MAANAOUI
Vector orthogonal polynomials.
J. Comput. Appl. Math., 32 (1990) 59-68.

[66] R.P. EDDY
Extrapolating to the limit of a vector sequence.
in *"Information Linkage between Applied Mathematicians and Industry"*, P.C.C. Wang ed., Academic Press, New-York, pp. 387-396.

[67] S.C. EISENSTAT, H.C. ELMANN, M. SCHULTZ
Variational iterative methods for nonsymmetric systems of linear equations.
SIAM J. Numer. Anal., 20 (1983), 345-357.

[68] J. FAVARD
Sur les polynômes de Tchebicheff.
C.R. Acad. Sci. Paris, 200 (1935), 2052-2053.

[69] M. FIEDLER
Special matrices and their applications in numerical mathematics.
M. Nijhoff, Dordrecht, 1986.

[70] R. FLETCHER
Conjugate gradient methods for indefinite systems.
in *"Numerical Analysis. Dundee 1975"* , LNM 506, Springer-Verlag,
Berlin, 1976, pp. 73-89.

[71] W.F. FORD, A. SIDI
*An algorithm for a generalization of the Richardson extrapolation
process.*
SIAM J. Numer. Anal., 24 (1987), 1212-1232.

[72] W.F. FORD, A. SIDI
Recursive algorithms for vector extrapolation methods.
Appl. Numer. Math., 4 (1988), 477-489.

[73] C. FOURGEAUD, A. FUCHS
Statistique.
Dunod, Paris, 1972.

[74] M. GASCA
Identidades para determinantes y eliminacion.
Actas VII CEDYA, Granada, 1984, pp. 163-167.

[75] M. GASCA, E. LEBRON
On Aitken-Neville formulae for multivariate interpolation.
in *"Numerical Approximation of Partial Differential Equation"*, E.L.
Ortiz ed., Elsevier, Amsterdam, 1987.

[76] M. GASCA, E. LEBRON
Elimination techniques and interpolation.
J. Comput. Appl. Math., 19 (1987), 125-132.

[77] M. GASCA, G. MÜHLBACH
Generalized Schur-complements and a test for total positivity.
Appl. Numer. Math., 3 (1987), 215-232.

[78] W. GAUTSCHI
A survey of Gauss-Christoffel quadrature formulae.
in *"E.B. Christoffel"*, P.L. Butzer and F. Fehér eds., Birkhäuser-Verlag,
Basel, 1981, pp. 72-147.

[79] B. GERMAIN-BONNE
Estimation de la limite de suites et formalisation de procédés d'accélération de la convergence.
Thèse d'Etat, Université de Lille 1, 1978.

[80] J. GILEWICZ
Numerical detection of the best Padé approximant and determination of the Fourier coefficients of insufficiently sampled functions.
in *"Padé Approximants and their Applications"*, P.R. Graves-Morris ed., Academic Press, New-York, 1973, pp. 99-103.

[81] P. GONZALEZ VERA
Two point Padé-type approximants : a new algebraic approach.
Publication ANO-157, Université de Lille 1, 1986.

[82] E. GOURSAT
Analyse.
Gauthier-Villars, Paris, 1915.

[83] W.B. GRAGG
Matrix interpretations and applications of the continued fraction algorithm.
Rocky Mt. J. Math., 4 (1974), 213-225.

[84] H.L. GRAY, T.A. ATCHISON, G.V. McWILLIAMS
Higher order G transformations.
SIAM J. Numer. Anal., 8 (1971), 365-381.

[85] I. HACCART
Accélération de la convergence de suites, séries et intégrales doubles.
Thèse 3ème cycle, Université de Lille 1, 1983.

[86] G. HALL, J.M. WATT, eds.
Modern numerical methods for ordinary differential equations.
Clarendon Press, Oxford, 1976.

[87] T. HAVIE
Generalized Neville type extrapolation schemes.
BIT, 19 (1979), 204-213.

[88] T. HAVIE
Remarks on a unified theory for classical and generalized interpolation and extrapolation.
BIT, 21 (1981), 465-474.

[89] T. HAVIE
An algorithm for iterative interpolation/extrapolation using generalized rational functions.
Math. and Comp., 4/82, ISBN 82-7151-047-9, University of Trondheim, Norway, 1982.

[90] T. HAVIE
Aitken/Neville type algorithms for iterative interpolation /extrapolation using generalized rational functions.
Math. and Comp., 5/82, ISBN 82-7151-048-7, University of Trondheim, Norway, 1982.

[91] T. HAVIE
Two algorithms for iterative interpolation and extrapolation using generalized rational functions.
Math. and Comp., 4/83, ISBN 82-7151-056-8, University of Trondheim, Norway, 1983.

[92] T. HAVIE
Remarks on some generalizations of the E-algorithm to vector spaces.
Math. and Comp., 2/84, ISBN 82-7151-058-4, University of Trondheim, Norway, 1984.

[93] T. HAVIE, M.J.D. POWELL
An application of Gaussian elimination to interpolation by generalized rational functions.
in *"Rational Approximation and Interpolation"*, P.R. Graves-Morris et al. eds., LNM 1105, Springer-Verlag, Berlin, 1984, pp. 442-452.

[94] C.J. HEGEDÜS
Newton's recursive interpolation in R^n.
in *"Numerical Methods"*, Colloquia Mathematica Societatis Janos Bolyai, vol. 50, Miskolc, 1986, pp. 605-623.

[95] **E. HENDRIKSEN**
Private communication, september 1989.

[96] **E. HENDRIKSEN, O. NJÅSTAD**
Biorthogonal Laurent polynomials with biorthogonal derivatives.
Preprint, 1988.

[97] **E. HENDRIKSEN, H. VAN ROSSUM**
Moment methods in Padé approximation.
J. Approximation Theory, 35 (1982), 250-263.

[98] **P. HENRICI**
Elements of numerical analysis.
J. Wiley, New-York, 1964.

[99] **F.B. HILDEBRAND**
Introduction to numerical analysis.
Mc Graw-Hill, New-York, 1956.

[100] **J. VAN ISEGHEM**
Padé-type approximants of exp(-z) *whose denominators are* $(1+z/n)^n$.
Numer. Math., 43 (1984), 283-292.

[101] **J. VAN ISEGHEM**
Vector Padé approximants.
in *"Numerical Mathematics and Applications"*, R. Vichnevetsky and J. Vignes eds., North-Holland, Amsterdam, 1986, pp. 73-77.

[102] **J. VAN ISEGHEM**
An extended cross rule for vector Padé approximants.
Appl. Numer. Math., 2 (1986), 143-155.

[103] **J. VAN ISEGHEM**
Vector orthogonal relations. Vector QD-algorithm.
J. Comput. Appl. Math., 19 (1987), 141-150.

[104] **J. VAN ISEGHEM**
Laplace transform inversion and Padé-type approximants.
Appl. Numer. Math., 3 (1987), 529-538.

[105] **J. VAN ISEGHEM**
Approximants de Padé vectoriels.
Thèse d'Etat, Université de Lille 1, 1987.

[106] A. ISERLES, P.E. KOCH, S.P. NØRSETT, J.M. SANZ-SERNA
Orthogonality and approximation in a Sobolev space.
Preprint, 1989.

[107] A. ISERLES, S.P. NØRSETT,
Bi-orthogonal polynomials.
in *"Polynômes Orthogonaux et Applications"*, C. Brezinski et al. eds.,
LNM 1171, Springer-Verlag, Berlin, 1985, pp. 92-100.

[108] A. ISERLES, S.P. NØRSETT,
Two-step methods and bi-orthogonality.
Math. Comput., 49 (1987), 543-552.

[109] A. ISERLES, S.P. NØRSETT
Bi-orthogonality and zeros of transformed polynomials.
J. Comput. Appl. Math, 19 (1987), 39-45.

[110] A. ISERLES, S.P. NØRSETT
On the theory of biorthogonal polynomials.
Trans. Am. Math. Soc., 306 (1988), 455-474.

[111] A. ISERLES, S.P. NØRSETT
Christoffel-Darboux-type formulae and a recurrence for bi-orthogonal polynomials.
Constr. Approx., 5 (1989), 437-453.

[112] A. ISERLES, E.B. SAFF
Biorthogonality in rational approximation.
J. Comput. Appl. Math., 19 (1987), 47-54.

[113] K. JBILOU
Méthodes d'extrapolation et de projection. Applications aux suites de vecteurs.
Thèse 3ème cycle, Université de Lille 1, 1988.

[114] K. JBILOU, H. SADOK
Some results about vector extrapolation methods and related fixed point iterations.
J. Comput. Appl. Math., to appear.

[115] W. B. JONES, W.J. THRON
Survey of continued fraction methods of solving moment problems and related topics.
in *"Analytic Theory of Continued Fractions"*, W.B. Jones et al. eds.,
LNM 932, Springer-Verlag, Berlin, 1982, pp. 4-37.

[116] D.S. KIM
A combinatorial approach to biorthogonal polynomials.
Ph. D., University of Minnesota at Minneapolis, 1989.

[117] J.D.E. KONHAUSER
Some properties of biorthogonal polynomials.
J. Math. Anal. Appl., 11 (1965), 242-260.

[118] A.S. KRONROD
Nodes and weights of quadrature formulas.
Consultants Bureau, New-York, 1965.

[119] F.M. LARKIN
Optimal approximation in Hilbert spaces with reproducing kernel functions.
Math. Comput., 24 (1970), 911-921.

[120] F.M. LARKIN
Gaussian measure in Hilbert space and applications in numerical anlaysis.
Rocky Mt. J. Math., 2 (1972), 379-421.

[121] A. LEMBARKI
Méthodes de projection et extensions : étude théorique et pratique.
Thèse 3ème cycle, Université de Lille 1, 1984.

[122] A. LE MEHAUTE
Interpolation d'Hermite itérée.
in *"Informatique et Calcul"*, P. Chenin et al. eds., Masson, Paris, 1986, pp. 77-81.

[123] P. LINZ
Theoretical numerical analysis.
J. Wiley, New-York, 1979.

[124] S.L. LOI
Quadratic approximation and its application to acceleration of convergence.
Ph. D., University of Canterbury, New Zealand, 1982.

[125] S.L. LOI
A general algorithm for rational interpolation.
Research Report 33, University of Canterbury, New Zealand, 1984.

[126] **S.L. LOI, A.W. McINNES**
An algorithm for generalized rational interpolation..
BIT, 23 (1983), 105-117.

[127] **S.L. LOI, A.W. McINNES**
An algorithm for the quadratic approximation
J. Comput. Appl. Math., 11 (1984), 161-174.

[128] **I.M. LONGMAN**
The application of rational approximations to the solution of problems in theoretical seismology.
Bull. Seismol. Soc. Amer., 56 (1966), 1045-1065.

[129] **I.M. LONGMAN, M. SHARIR**
Laplace transform inversion of rational functions.
J. Astr. Soc., 25 (1971), 299-305.

[130] **T. LYCHE**
A recurrence relation for Chebyshevian B-splines.
Constr. Approx., 1 (1985), 155-173.

[131] **P. MARONI**
Généralisation du théorème de Shohat-Favard sur les polynômes orthogonaux.
C.R. Acad. Sci. Paris, sér. I, 293 (1981), 19-22.

[132] **P. MARONI**
L'orthogonalité et les récurrences d'ordre supérieur à deux.
Ann. Fac. Sci. Toulouse, 10 (1989), 105.

[133] **G. MEINARDUS, G.D. TAYLOR**
Lower estimates for the error of best uniform approximation.
J. Approximation Theory, 16 (1976), 150-161.

[134] **J. MEINGUET**
Multivariate interpolation at arbitrary points made simple.
Z. Angew. Math. Phys., 30 (1979), 292-304.

[135] **M. MESINA**
Convergence acceleration for the iterative solution of the equation x = Ax + f.
Comput. Math. Appl. Mech. Engrg., 10 (1977), 165-173.

[136] W.L. MIRANKER
Numerical methods for stiff equations.
D. Reidel, Dordrecht, 1981.

[137] G. MONEGATO
Stieltjes polynomials and related quadrature rules.
SIAM Rev., 24 (1982), 137-158.

[138] M. MORANDI CECCHI, M. REDIVO ZAGLIA
*Mathematical programming techniques to solve biharmonic
problems by a recursive projection algorithm.*
J. Comput. Appl. Math., to appear.

[139] G. MÜHLBACH
*A recurrence formula for generalized divided differences and some
applications.*
J. Approximation Theory, 9 (1973), 165-172.

[140] G. MÜHLBACH
*Neville-Aitken algorithms for interpolation by functions of Cebysev-
systems in the sense of Newton and in a generalized sense of
Hermite .*
in "Theory of Approximation with Applications", A.G. Law and B.N.
Sahney eds., Academic Press, New-York, 1976, pp. 200-212.

[141] G. MÜHLBACH
The general Neville-Aitken algorithm and some applications.
Numer. Math., 31 (1978), 97-110.

[142] G. MÜHLBACH
*Extrapolation algorithms as elimination techniques with applications
to systems of equations.*
Report 152, University of Hannover, 1982.

[143] G. MÜHLBACH
Algorithmes d'extrapolation.
Publication ANO-118, Université de Lille 1, 1984.

[144] G. MÜHLBACH
*Two composition methods for solving certain systems of linear
equations.*
Numer. Math., 46 (1985), 339-349.

[145] G. MÜHLBACH
Linear and quasilinear extrapolation algorithms.
in *"Numerical Mathematics and Applications"*, R. Vichnevetsky and J. Vignes eds., North-Holland, Amsterdam, 1986, pp. 65-71.

[146] G. MÜHLBACH
On multivariate interpolation by generalized polynomials on subsets of grids.
Computing, 40 (1988), 201-215.

[147] G. MÜHLBACH
On interpolation by generalized polynomials of one and of several variables.
Det. Kong. Norske Vid. Selsk. Skr., 2 (1989) 87-101.

[148] G. MÜHLBACH, M. GASCA
A generalisation of Sylvester's identity on determinants and some applications.
Linear Algebra Appl., 66 (1985), 221-234.

[149] J. NUTTALL
Convergence of Padé approximants for the Bethe-Salpeter amplitude.
Phys. Rev., 157 (1967), 1312-1316.

[150] D.V. OUELLETTE
Schur complements and statistics.
Linear Algebra Appl., 36 (1981), 187-295.

[151] E. PARZEN
An approach to time series analysis.
Ann. Math. Stat., 32 (1961), 951-989.

[152] E. PARZEN
Probability density functionals and reproducing kernel Hilbert spaces.
in *"Time Series Analysis"*, M. Rosenblatt ed., J. Wiley, New-York, 1963, pp. 155-169.

[153] S. PASZKOWSKI
Recurrence relations in Padé-Hermite approximation.
J. Comput. Appl. Math., 19 (1987), 99-107.

[154] J.L. PHILLIPS
The use of collocation as a projection method for solving linear operator equations.
SIAM J. Numer. Anal., 9 (1972), 14-28.

[155] M. PREVOST
Sommation de certaines séries formelles par approximation de la fonction génératrice.
Thèse 3ème cycle, Université de Lille 1, 1983.

[156] M. PREVOST
Stieltjes- and Geronimus-type polynomials.
J. Comput. Appl. Mat., 21 (1988), 133-144.

[157] M. PREVOST
Determinantal expression for partial Padé approximants.
Appl. Numer. Math., 6 (1989/1990), 221-224.

[158] M. PREVOST
Private communication, may 1989.

[159] B.P. PUGACHEV
Acceleration of the convergence of iterative processes and a method of solving systems of non-linear equations.
USSR Comput. Maths. Math. Phys., 17 (1978), 199-207.

[160] W.C. PYE, T.A. ATCHISON
An algorithm for the computation of higher order G-transformation.
SIAM J. Numer. Anal., 10 (1973), 1-7.

[161] S.M. ROMAN, G.C. ROTA
The umbral calculus.
Adv. Math., 27 (1978), 95-188.

[162] H. VAN ROSSUM
A theory of orthogonal polynomials based on the Padé table.
Thesis, University of Utrecht, Van Gorcum, Assen, 1953.

[163] H. VAN ROSSUM
Generalized Padé approximants.
in "Approximation Theory III", E.W. Cheney ed., Academic Press, New-York, 1980.

[164] H. VAN ROSSUM
Formally biorthogonal polynomials.
in *"Padé Approximation and its Applications. Amsterdam 1980"*,
M.G. de Bruin and H. Van Rossum eds., LNM 888, Springer-Verlag,
Berlin, 1981, pp. 341-351.

[165] Y. SAAD
*Krylov subspace methods for solving large unsymmetric linear
systems*
Math. Comput., 37 (1981), 105-126.

[166] H. SADOK
*Accélération de la convergence de suites vectorielles et méthodes de
point fixe.*
Thèse, Université de Lille 1, 1988.

[167] H. SADOK
About Henrici's transformation for accelerating vector sequences.
J. Comput. Appl. Math., 29 (1990), 101-110.

[168] C. SCHNEIDER
*Vereinfachte Rekursionen zur Richardson-Extrapolation in Spe-
cialfällen.*
Numer. Math., 24 (1975), 177-184.

[169] R.E. SHAFER
On quadratic approximation.
SIAM J. Numer. Anal., 11 (1974), 447-460.

[170] D. SHANKS
*Non linear transformations of divergent and slowly convergent
sequences.*
J. Math., Phys., 34 (1955), 1-42.

[171] B. SHEKHTMAN
Interpolation in abstract spaces.
Ph. D., Kent State University, 1980.

[172] J. SHOHAT
Sur les polynômes orthogonaux généralisés.
C.R. Acad. Sci. Paris, 207 (1938), 556-558.

[173] A. SIDI
Convergence and stability properties of minimal polynomial and reduced rank extrapolation algorithms.
SIAM J. Numer. Anal., 23 (1986), 197-209.

[174] A. SIDI
Extrapolation vs. projection methods for linear systems of equations.
J. Comput. Appl. Math., 22 (1988), 71-88.

[175] A. SIDI, J. BRIDGER
Convergence and stability analysis for some vector extrapolation methods in the presence of defective iteration matrices.
J. Comput. Appl. Math., 22 (1988), 35-61.

[176] A. SIDI, W.F. FORD, D.A. SMITH
Acceleration of convergence of vector sequences.
SIAM J. Numer. Anal., 23 (1986), 178-196.

[177] I. SINGER
Bases in Banach Spaces, I.
Springer-Verlag, Berlin, 1970.

[178] F. SLOBODA
Nonlinear iterative methods and parallel computation.
Apl. Mat., 21 (1975), 252-262.

[179] F. SLOBODA
A parallel projection method for linear algebraic systems.
Apl., Mat., 23 (1978), 185-198.

[180] D.A. SMITH, W.F. FORD, A. SIDI
Extrapolation methods for vector sequences.
SIAM Rev., 29 (1987), 199-233.

[181] D.A. SMITH, W.F. FORD, A. SIDI
Correction to "Extrapolation methods for vector sequences".
SIAM Rev., 30 (1988), 623-624.

[182] R.C.E. TAN
Computing derivatives of eigensystems by the topological ε-algorithm.
Appl. Numer. Math., 3 (1987), 539-550.

[183] R.C.E. TAN
Implementation of the topological ε-algorihtm.
SIAM J. Sci. Stat. Comput., 9 (1988), 839-848.

[184] B. VERDONK
Different techniques for multivariate rational interpolation and Padé approximation.
Thesis, University of Antwerp, 1988.

[185] G. VIENNOT
Une théorie combinatoire des polynômes orthogonaux généraux.
Université de Québec, Montréal, 1983.

[186] Yu. V. VOROBYEV
Method of moments in applied mathematics.
Gordon and Breach, New-York, 1965.

[187] H.L. WEINERT, ed.
Reproducing kernel Hilbert spaces.
Hutchinson Ross Publ. Co., Stroudsburg, 1982.

[188] J. WIMP
Sequence transformations and their applications.
Acadamic Press, New-York, 1981.

[189] P. WYNN
On a device for computing the e_m (S_n) transformation.
MTAC, 10 (1956), 91-96.

[190] P. WYNN
On a procrustean technique for the numerical transformation of slowly convergent sequences and series.
Proc. Cambridge Phil. Soc., 52 (1956), 663-671.

[191] P. WYNN
Upon systems of recursions which obtain among the quotients of the Padé table.
Numer. Math., 8 (1966), 264-269.

[192] P. WYNN
Upon some continuous prediction algorithms.
Calcolo, 9 (1972), 197-234 ; 235-278.

Index